KU-340-927

INDUSTRIAL RISK MANAGEMENT:

A LIFE-CYCLE ENGINEERING APPROACH

Industrial Risk Management: A Life-Cycle Engineering Approach

A selection of papers presented at the International Conference on Industrial Risk Management at the Swiss Federal Institute of Technology, Zürich, Switzerland, 16–17 January 1989

edited by

Thomas Bernold

Istituto Dalle Molle di Metodologie Interdisciplinari, Lugano, Switzerland

on behalf of

the Gottlieb Duttweiler Institute, Ruschlikon, Switzerland

This material has also been published as a special issue of the *Journal of Occupational Accidents*, Vol. 13 (1 & 2)

ELSEVIER

AMSTERDAM — OXFORD — NEW YORK — TOKYO 1990

ELSEVIER SCIENCE PUBLISHERS B.V.
Sara Burgerhartstraat 25
P.O. Box 211, 1000 AE Amsterdam, The Netherlands

Distributors for the United States and Canada:

ELSEVIER SCIENCE PUBLISHING COMPANY INC.
655, Avenue of the Americas
New York, NY 10010, U.S.A.

Library of Congress Cataloging-in-Publication Data

International Conference on Industrial Risk Management (199 : Zurich, Switzerland)
Industrial risk management : a life-cycle engineering approach : a selection of papers presented at the International Conference on Industrial Risk Management, Zürich, Switzerland, 16-17 January 1989 at the Swiss Federal Institute of Technology / edited by Thomas Bernold.
p. cm.
"This material has also been published as a special issue of the Journal of ocupational accidents, vol. 13 (1&2)."
ISBN 0-444-88004-6 (alk. paper)
1. Reliability (Engineering)--Congresses. 2. Product life cycles--Congresses. 3. Risk management--Congresses. I. Bernold, Thomas, 1951- . II. Ecole polytechnique fédérale de Lausanne. III. Title.
TS173.I57 1989
620'.00452--dc20 90-14032
CIP

ISBN 0-444-88004-6

This book is printed on acid-free paper.

Printed in The Netherlands

Preface

Product-life: from design to disposal

Product-life is a new approach to technical product safety that addresses the entire life of a product, from its conception through its development, production, distribution and final disposal. Until now, safety and risk have been assessed independently in some stages of a product's existence. *Product-life* presents a comprehensive and integrated approach to risk management and industrial environmental policy and thus provides for the reduction of potential risk to both human and the environment.

The need for an improved product risk management approach is highlighted by the frequency of industrial accidents and the increasing potential severity of the effects of these accidents. While specific safety systems may have become more reliable, technological innovation has resulted in greater risks and the development of safety systems has not kept pace with the increased risk. The potential effects of systems failure have grown considerably: consequences are now both local and global; effecting life that is exposed directly as well as future generations.

The complexity of today's products and systems and the increased environmental vulnerability indicate that current problems with risk management are not only quantitative, but qualitative: a rethinking of safety management design and management is necessary.

This international conference of 16 and 17 January 1989 was sponsored by the ETH Zürich (Swiss Federal Institute of Technology), the Geneva Association, and the Gottlieb Duttweiler Institute. It focuses on a comprehensive approach to risk assessment and management for products. This approach is summarized by the formula: *Risk prevention is best achieved through the synthesis of knowledge about a product throughout its entire life.* The keynotes develop the topic from the various perspectives starting with risk- and loss analysis, human factors, economy of safety, liability and compensation, risk assessment, management and prevention as well as the role of innovation to overcome the challenges of a civilisation which depends increasingly on complex technical systems.

Case studies from different industries ranging from pharmaceuticals to high tech investment goods illustrate the specific approaches to apply this concept in current safety management situations. The case studies emphasize the crossdisciplinary integration of information that must occur at every stage of a product's life so as to assure safety. The presentations address comprehensive strategies for industry and services and review the potential benefits of these strategies.

We should, however, not forget that human beings have a tendency to become more careless as the technical systems are designed for more intrinsic safety. Human error, more than once, was at the core of many recent accidents and catastrophes. It seems that we have yet to persue a long way before the safety of the overall system of man and machine reaches a satisfactory stage. This, however, reaches far beyond technical design issues.

T. BERNOLD
Zürich, January 1990

CONTENTS

INTRODUCTION

The need for friendly plants*

Trevor A. Kletz

University of Technology, Loughborough, U.K.

ABSTRACT

Kletz, T.A., 1990. The need for friendly plants. *Journal of Occupational Accidents,* 13: 3–13.

Plants should be designed as far as possible so that they are *user-friendly*, that is, so that errors by operators or maintenance workers or equipment failure does not have serious effects on safety, output or efficiency. Some examples are described. Thus friendly plants contain small or zero inventories of hazardous materials, are simple, with few opportunities for error, easy to control and hard to assemble incorrectly. The action needed for the design of friendly plants is discussed.

INTRODUCTION

In both the process and nuclear industries errors by operators and maintenance workers are recognised as a major cause of accidents and the importance of training, auditing, inspection and motivation has often been emphasised. However, it is difficult for operators and maintenance workers to keep up an error-free performance all day, every day. Designers have a second chance, opportunities to go over their designs again, but not operators and maintenance workers. Other accidents are due to the failure of components. Plants should therefore be designed, whenever possible, so that they are *user-friendly*, to borrow a computer term, so that they can tolerate departures from ideal performance–by operators, maintenance workers or equipment–without serious effects on safety output or efficiency.

Some of the characteristics of friendly plants are summarised in Table 1 and are discussed in detail below. Thus friendly plants contain low inventories of hazardous materials or safer materials instead; they are simpler, with fewer leakage points or opportunities for error; they are less dependent on added-on

---

*Presented at the International Conference on Industrial Risk Management, Zürich, Switzerland, 16-17 January 1989.

**This paper was originally presented at an American Institute of Chemical Engineers (AIChE) meeting in August 1988 and published, with amendments, in *Chemical Engineering Progress*, Vol. 85, No. 7, July 1989, pp. 18–26.

TABLE 1

Characteristics of friendly plants with examples (for further details see text, and Kletz (1978, 1985a, 1987) and Wade et al. (1987)

	Characteristic		Friendliness	Hostility
1	Intensification	Distillation	Higee	Conventional
		Heat transfer	Miniaturized	Conventional
		Nitrogycerine manufacture	NAB process	Batch process
		Intermediate storage	Small or nil	Large
		Reaction	Vapour phase	Liquid phase
			Tubular reactor	Pot reactor
2	Substitution	Heat transfer media	Non-flammable	Flammable
		Solvents	Non-flammable	Flammable
		Chlorine manufacture	Membrane cells	Hg and asbestos cells
		Carbaryl production	Israeli process	Bhopal process
3	Attenuation	Liquefied gases	Refrigerated	Under pressure
		Explosive powders	Slurried	Dry
		Runaway reactants	Diluted	Neat
		Any material	Vapour	Liquid
4	Simpler design with fewer leakage points or opportunities for error		Hazards avoided	Hazards controlled by added equipment
			Single stream	Multi-stream with many cross-overs
			Dedicated plant	Multipurpose plant
			One big plant	Many small plants
		Spares	Uninstalled	Installed
5	No knock-on effects		Open construction	Closed buildings
			Fire-breaks	No fire-breaks
		Tank roof	Weak seam	Strong seam
		Horizontal cylinder	Pointing away from other equipment	Pointing at other equipment
6	Incorrect assembly impossible	Compressor valves	Non-interchangeable	Interchangeable
		Device for adding water to oil	Cannot point upstream	Can point upstream
7	Status clear		Rising spindle valve or ball valve with fixed handle	Non-rising spindle valve
			Spectacle plate[a]	Slip-plate
8	Tolerant of maloperation or poor maintenance		Continuous plant	Batch plant
			Spiral wound gasket	CAF gasket
			Expansion loop	Bellows
			Fixed pipe	Hose
			Articulated arm	Hose
			Bolted joint	Quick-release coupling
			Metal	Glass, plastic

TABLE 1 (Continued)

	Characteristic		Friendliness	Hostility
9	Low leak rate		Spiral wound gasket	CAF gasket
			Tubular reactor	Pot reactor
			Vapour phase reactor	Liquid phase reactor
10	Easier to control	Response to change	Flat	Steep
		Negative temp. coeff.	Processes in which rise in T produces reaction stopper	Most processes
			Most nuclear reactors	Chernobyl reactor
		Slow response	AGR	PWR
		Less dependant on added on safety systems	AGR, FBR, HTGR PIUS	PWR
11	Software	Errors easy to detect and correct	Some PES	Some PES
		Training and instructions	Some	Most
		Gaskets, nuts, bolts etc.	Few types stocked	Many types stocked
12	Other industries	Continuous movement[b]	Rotating engine	Reciprocating engine
		Helicopters with 2 rotors	Cannot touch	Can touch
		Chloroform dispenser	Reverse connection possible	Reverse connection impossible
13	Analogies		Lamb	Lion
			Bungalow	Staircase
			Tricycle	Bicycle
		Marble on saucer	concave up	convex up
		Boiled egg	-pointed end up	-blunt end up
			-hard-boiled	-soft-boiled
			-medieval egg-cup	-standard egg-cup

[a]A spectacle plate is easier to fit (in rigid piping) and easier to find.
[b]In practice reciprocating internal combustion engines are not less friendly than rotating engines though one might expect that equipment which continually starts and stops would be less reliable.
AGR = Advanced gas-cooled reactor
HTGR = High temperature gas reactor
PIUS = Process inherent ultimate safety reactor
FBR = Fast breeder reactor
PES = Programmable electronic system
PWR = Pressurised water reactor

safety systems which may be neglected or ignored or may fail; they respond more slowly and less steeply to changes. Those errors that are made or failures that do occur do not produce extensive knock-on effects. Thus spiral wound gaskets are more friendly than caf gaskets, expansion loops than bellows, articulated arms than hoses, bolted joints than quick-release couplings, as they are more tolerant of maintenance errors or ill-treatment. Rising-spindle valves are friendlier than valves with non-rising spindles, and spectacle plates are friendlier than slip-plates, as their positions are less likely to be misread.

It is the theme of this paper that instead of designing plants, identifying hazards and adding on equipment to control the hazards, or expecting operators to control them, we should make more effort to choose basic designs, and design details, that are user-friendly. Friendly design is thus an extension of the concept of inherently safer design that the author has advocated for a number of years (Kletz, 1978, 1985a, 1987).

1. INTENSIFICATION

One obvious way of making plants safer is to reduce inventories of hazardous materials in process and storage. Little conscious thought was, however, given to this subject by most chemical engineers until after the explosion at Flixborough in 1974. Until then it was usual to design a plant and accept whatever inventory was called for by the design, controlling hazards by adding on protective equipment such as trips and alarms, fire-protection, fire-fighting. In fact, a great deal can be done to reduce inventories in reaction, distillation, heat transfer and other unit operations and many examples are described by Kletz (1978, 1985a, 1987). At Bhopal (1984) the material which leaked, killing about 2000 people, was an intermediate which it was convenient but not essential to store. As a result many companies have now reduced their stocks of hazardous intermediates and there has been renewed interest in intensification, as it is called (Wade et al., 1987).

There is, of course, no advantage in reducing the inventory in a plant by building several small plants instead of one big one. This merely increases the number of joints, pumps, etc. from which hazardous materials can leak without reducing the total inventory. We should use our skill as engineers to design small plants with the same output as large ones.

2. SUBSTITUTION

If intensification is not possible, then an alternative is substitution, using a safer material in place of a hazardous one. Thus it is possible to replace flammable refrigerants and heat transfer media by non-flammable ones, hazardous products by safer ones, processes which use hazardous raw materials or intermediates by processes which do not. There are many examples in the papers already cited.

Intensification, when it is practicable, is better than substitution as it brings about greater reductions in cost. If less material is present we need smaller pipes and vessels, smaller structures and foundations. Much of the pressure for intensification has come from those who are primarily concerned with cost reduction. In fact, friendliness in plant design is not just an isolated but desirable concept but part of a total package of measures, including cost reduction, lower energy usage and simplification that the chemical industry needs to adopt in the years ahead.

3. ATTENUATION

Another alternative to intensification is attenuation, using a hazardous material under the least hazardous conditions. Thus liquefied chlorine and ammonia can be stored as refrigerated liquids at atmospheric pressure instead of storing them under pressure at ambient temperature. Dyestuffs which form explosive dusts can be handled as slurries. There are other examples in the references already cited.

4. SIMPLICITY

Simpler plants are friendlier as they provide fewer opportunities for error and less equipment which can go wrong. Some of the reasons for complication in plant design are (Kletz, 1985a, Chaps. 5 and 6):

- The need to add on complex equipment to control hazards. If we can intensify, substitute or attenuate, as already discussed, we need less added-on protective equipment and plants will therefore be simpler.
- A desire for flexibility. Multi-stream plants with numerous crossovers and valves, so that any item can be used on any stream, have numerous leakage points and errors in valve settings are easy.
- Lavish provision of installed spares with the accompanying isolation and change-over valves.
- Continuing to follow rules or practices which are no longer necessary.
- Design procedures which result in a failure to identify hazards until late in design. By this time it is impossible to avoid the hazard and all we can do is add on complex equipment to control it (for example, see Kletz, 1985a).

5. KNOCK-ON EFFECTS

Friendly plants are designed so that those incidents that do occur do not produce knock-on or domino effects. For example:

- Friendly plants are provided with fire-breaks, about 15 m wide, between sections, like fire-breaks in a forest, to restrict the spread of fire.
- When flammable materials are handled, friendly plants are built out-of-doors so that leaks can be dispersed by natural ventilation. Indoors a few tens of kilograms are sufficient for an explosion that can destroy the building. Out-of-doors a few tonnes are necessary for serious damage. A roof over equipment such as compressors is acceptable but walls should be avoided.
- Storage tanks are normally built so that the roof-wall weld will fail before

the base-wall weld, thus preventing spillage of the contents. In general, in designing equipment we should consider the way in which it is most likely to fail and, when possible, locate or design the equipment so as to minimise the consequences.

6. INCORRECT ASSEMBLY IMPOSSIBLE

Friendly plants are designed so that incorrect assembly is difficult or impossible. For example, compressor valves should be designed so that inlet and exit valves cannot be interchanged.

Another example is shown in Fig. 1. A device was designed for adding an aqueous stream into the centre of an oil stream in such a way that mixing would occur and corrosion would be minimised (b). It was assembled as shown in (c) and corrosion was worse. Once it was assembled it was impossible to see that it had been assembled incorrectly. It should have been designed so that it could not be assembled wrongly, or at least so that any wrong assembly was apparent.

7. STATUS CLEAR

With friendly equipment it is possible to see at a glance if it has been assembled or installed incorrectly or whether it is in the open or shut position. One example has just been quoted. Other examples are:

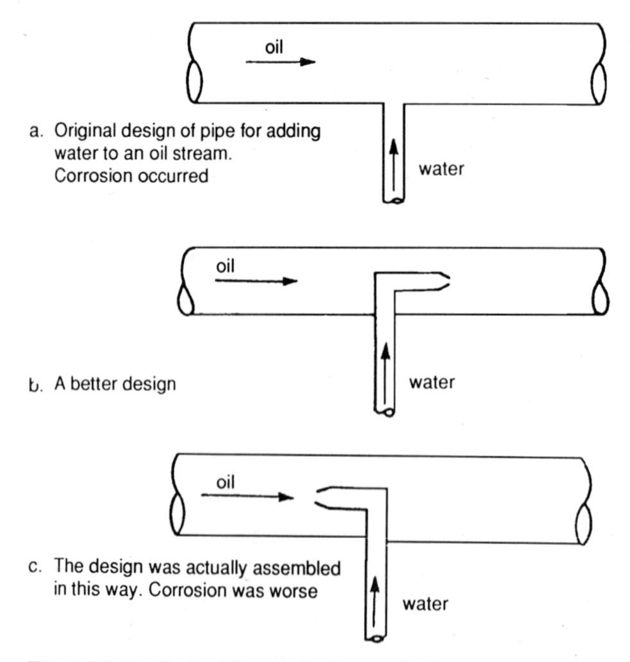

Fig. 1. Methods of adding water to an oil stream.

- Non-return (check) valves should be marked so that installation the wrong way round is obvious. It should not be necessary to look for a faint arrow hardly visible beneath the dirt.
- Gate valves with rising spindles are friendlier than valves with nonrising spindles, as it is easy to see whether they are open or shut. Ball valves are friendly if the handles cannot be replaced in the wrong position.
- Spectacle (figure 8) plates are friendlier than slip-plates (spades) as their position is apparent at a glance. If slip-plates are used their projecting tags should be readily visible, even when the line is insulated. In addition, spectacle plates are easier to fit than slip-plates, if the piping is rigid, and are always available on the job. It is not necessary to search for one, as with slip-plates.

8. TOLERANCE

Friendly equipment will tolerate poor installation or operation without failure. Thus spiral wound gaskets are friendlier than plain gaskets as if the bolts work loose, or are not tightened correctly, the leak rate is much less. Expansion loops in pipework are more tolerant of poor installation than bellows. Fixed pipes, or articulated arms, if flexibility is necessary, are friendlier than hoses. For most applications, metal is friendlier than glass or plastic.

Bolted joints are friendlier than quick-release couplings. The former are usually dismantled by a fitter after issue of a permit-to-work. One man prepares the equipment and another man opens it up; the issue of the permit provides an opportunity to check that the correct precautions have been taken. In addition, if the joints are unbolted correctly, any trapped pressure is immediately apparent and the joint can be remade or the pressure allowed to blow off. In contrast many accidents have occurred because operators opened up equipment which was under pressure, without independent consideration of the hazards, using quick-release couplings. There are, however, designs of quick-release couplings which give the operator a second chance (Kletz, 1985b).

9. LOW LEAK-RATE

If friendly equipment does leak, it does so at a low rate which is easy to stop or control. Spiral-wound gaskets have already been mentioned. A tubular reactor is friendlier than a pot reactor. If the tube is narrow the leak rate is limited by the cross-section of the pipe and can be stopped by closing a valve in the pipe. Vapour phase reactors are friendlier than liquid phase reactors as the mass flow rate through a hole of a given size is less.

10. EASE OF CONTROL

Processes with a flat response to change are obviously friendlier than those with a steep response. Processes in which a rise of temperature decreases the rate of reaction are friendlier than those with a positive temperature coefficient, but this is a difficult ideal to achieve in the chemical industry. However, there are a few examples of processes in which a rise in temperature reduces the rate of reaction. For example, in the manufacture of peroxides water is removed by a dehydrating agent. If magnesium sulphate is used as the agent a rise in temperature causes release of water by the agent, diluting the reactants and stopping the reaction (Gerrirsen and van 't Land, 1985).

Nuclear reactors differ greatly in their ease of control and their dependence on added-on control and trip systems which may fail or may be neglected. The Chernobyl boiling water reactor (BWR) had a positive power coefficient, at low output rates (less than 20%). As the power output and temperature increased the rate of heat generation increased. The operators were told never to go below 20% output but they did not follow the instruction. There was no trip system to prevent them doing so. If there had been, the trip system could have failed or been disarmed. It is better to choose a design of reactor with a negative power coefficient. No other commercial design has a positive coefficient and the Chernobyl design has now been modified.

Pressurised water reactors (PWR) have negative power coefficients but are still dependant on added-on engineered cooling systems. If the normal cooling system fails then emergency systems are needed to prevent overheating. In contrast, advanced gas cooled reactors (AGR) are cooled to a substantial extent by convection if forced circulation is lost. Fast breeder reactors (FBR) and other designs still under development, such as the high temperature gas reactor (HTGR) and the PIUS (process inherent ultimate safety) reactor cannot overheat even if all cooling systems fail completely. In the PIUS design a water cooled reactor is immersed in a vessel containing borate solution. If coolant pressure is lost the reactor is flooded by the solution which stops the reaction and cools the reactor. No make-up water is needed for a week.

Gas cooled reactors also give the operator more time in which to react. Franklin writes, "When operators are subject to conditions of extreme urgency ... they will react in ways that lead to a high risk of promoting accidents rather than diminishing them. This is materially increased if operators are aware of the very small time margins that are available to them", and "It is much better to have reactors which, even if they do not secure the last few percent of capital cost effectiveness, provide the operator with half-an-hour to reflect on the consequences of the action before he needs to intervene (Franklin, 1986).

11. SOFTWARE

In some programmable electronic systems (PES) errors are much easier to detect and correct than in others. Using the terms *software* in the wider sense to cover all procedures, as distinct from hardware or equipment, some software is much friendlier than others. Training and instructions are obvious examples. Another example: If many types of gaskets or nuts and bolts are stocked, sooner or later the wrong type will be installed. It is better, and cheaper in the long run, to keep the number of types stocked to a minimum even though more expensive types than are strictly necessary are used for some applications.

12. OTHER INDUSTRIES

A helicopter crashed because the two rotors got out of phase and touched each other. The cause was said to be corrosion of a gear wheel and monitoring equipment was recommended. However, in a friendlier design the two rotors would have been located so that they could not touch each other.

One would have expected that reciprocating engines, which start and stop twice every cycle, would be less friendly and more troublesome than rotating

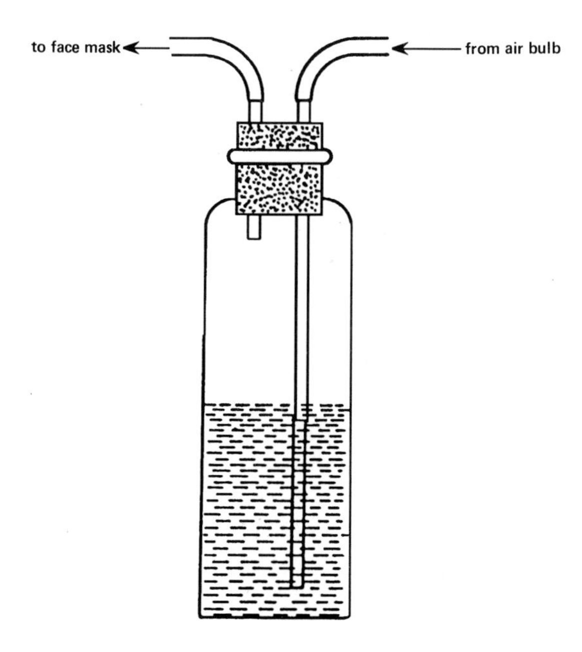

Fig. 2. Early chloroform dispenser.

engines but this does not seem to be the case. Though reciprocating steam engines have given way to turbines, the reciprocating internal combustion engine still reigns supreme.

In the early days of anaesthetics chloroform was mixed with air and piped to a facemask using the simple apparatus shown in Fig. 2, introduced in 1867. If the two pipes were interchanged liquid chloroform was supplied to the patient, with results that could be fatal. Redesigning the apparatus so that it was friendlier, so that the two pipes could not be interchanged was easy, but persuading doctors to use the new design was more difficult. They were reluctant to admit that they could make such a simple error. As late as 1928 the simple apparatus was still killing people (Sykes, 1960).

13. THE ROAD TO FRIENDLIER PLANTS

The following actions are suggested:

(1) Awareness of the problem. Designers need to be made aware that there is scope for improving the friendliness of the plants they design. This paper is intended as a step in that direction.

(2) Front end studies. To achieve some of the changes suggested, particularly those discussed in Sections 1–5, 9 and 10, it is necessary to carry out much more critical examination and systematic consideration of alternatives during the early stages of design than has been customary in most companies. Two studies are suggested. One at the conceptual stage when the process is being chosen and another at the flowsheet stage. For the latter the usual hazard and operability study (hazop) questions are suitable but with one qualification. In a normal hazop on a line diagram, if we are discussing *more of temperature* (say) we assume that it is undesirable and look for ways of preventing it. In a hazop of a flowsheet we should ask if *more of temperature* would be better. For the conceptual study some questions are suggested by Kletz (1985a, Chap. 4).

Many companies will say that they do consider alternatives during the early stages of plant design. However what is lacking in many companies is a formal, systematic, structured procedure of the hazop type.

Friendly plants are often cheaper than hostile ones.

If it costs $1 to fix a problem at the conceptual stage it will cost $10 at the flowsheet stage, $100 at the line diagram stage, $1000 after the plant is built and over $10,000 to clean up the mess after an accident.

(3) Detailed design studies. To achieve the detailed improvements suggested in Sections 6–8 and 11 it may be necessary to add a few questions to those asked during a normal hazop. For example, what types of valve, gasket etc. will be used?

14. ANALOGIES

Simple everyday analogies often help to explain ideas. The following are therefore suggested:

(1) Suppose the meat of lions was good to eat or their skins made good clothes. Our farmers would be asked to farm lions and they could do so though they would need cages round their fields instead of hedges. Only occasionally, as at Flixborough or Bhopal, would the lions break loose. But why keep lions when lambs will do instead?

(2) The most dangerous equipment in our homes is the stairs. More people are killed or injured on them than in any other way. The conventional solution is to train people to use the handrails, make sure the carpet is secure and keep the stairs free from junk. The inherently safer or friendlier solution is to buy a single story house (bungalow).

(3) A tricycle is friendlier than a bicycle.

(4) Control: it is easier to keep a marble on a concave-up saucer than on a convex-up saucer. Chernobyl (Section 10) was a marble on a convex surface.

(5) To prevent a boiled egg falling over and making a mess we can

- Hard boil it.
- Put it in the eggcup with the pointed end up so that the centre of gravity is lower.
- Use a design of eggcup in which the egg lies horizontal.

ACKNOWLEDGEMENTS

Thanks are due to the many colleagues, past and present, who have suggested ideas for this paper, to the Leverhulme Trust for financial support and to Wayne State University for permission to reproduce a paper originally presented there.

REFERENCES

Franklin, N., 1986. The accident at Chernobyl. Chem. Eng., 430: 17.

Gerrirsen, H.G. and van 't Land, C.M., 1985. Intrinsic continuous process safeguarding. I & EC Process Des. Devel., 24: 893.

Kletz, T.A., 1978. What you don't have, can't leak. Chem. & Ind., 6 May: 287.

Kletz, T.A., 1985a. Cheaper, Safer Plants. Institution of Chemical Engineers, 2nd edn. Third edition entitled Plant Design for Safety-A User-Friendly Approach, will be published later 1990, by Hemisphere Publishing Corporation.

Kletz, T.A., 1985b. What Went Wrong? Gulf Publishing Co., Sections 13.5 and 17.1

Kletz, T.A., 1987. Process Optimisation. Institutions of Chemical Engineers Symposium Series No. 100, p. 153.

Sykes, W.S., 1960. Essays on the First Hundred Years of Anaesthesia, Vol. 2. Churchill Livingstone, pp. 3-5.

Wade, D.E., Hendershot, D.C., Caputo, R.J. and Dale, S.E., 1987. In: Proc. International Symposium on Preventing Major Chemical Accidents. American Institute of Chemical Engineers, pp. 2.1, 3.1, 3.43 and 3.79.

FROM RISK ANALYSIS TO INNOVATION

Bridging engineering gaps and vacuums from design to disposal*

Ted Ferry

Edmonds, WA 98020, U.S.A.

ABSTRACT

Ferry, T., 1990. Bridging engineering gaps and vacuums from design to disposal. *Journal of Occupational Accidents*, 13: 17–31.

The best efforts of qualified personnel to engineer out system faults and ensure mishap-free performance systems are not successful. We find that aspects beyond our skills and resources preclude dealing with the risk prevention problems of both simple and complex systems and the interaction between non-connected systems. Risk analysis techniques, are typically unable to assess all signs and root causes analyses in a systems approach. A management-based overview approach is suggested to control the environmental forces that impinge on risk prevention, be they technical, human, social, political or organizational.

INTRODUCTION

This paper addresses gaps and vacuums in engineering approaches from design to disposal. It suggests ways to find, evaluate and fill these gaps and vacuums.

Initial design does not normally consider the idea leading to it. We seldom evaluate the concept itself, but instead seek to find the best engineering approach to a problem. Or, consider disposal. Do we know that disposal by burial or incineration leaves us with hazardous materials and waste problems? Conceptual blueprints should regard these problems in feasibility studies. Or, consider that nearly everything we now design is computer driven or related. Not all of us realize that software cost now often exceeds the hardware cost?

If conceptual action comes early in our thinking, what comes last, after disposal? It may well be litigation. If our professional approaches do not include complete risk assessments, then our litigious societies will haunt us, long after

*Presented at the International Conference on Industrial Risk Management, Zürich, Switzerland, 16–17 January 1989.

we finish the product or process. Banks now, or soon, will examine your risk performance record before financing new products, processes, and facilities. Possible litigation keeps many good products and processes from getting off-the-ground. Liability insurance needed to protect investors, designers, products, and processes can exceed all other combined startup costs. Litigation is only one gap in engineering approaches. Still others spring from new technologies.

NEW TECHNOLOGIES LEAVE MANY CONCERNS UNCHANGED

The technology explosion has not changed the human element. We are not skilled enough to write the human out of the system or handle it all by engineering and risk analysis approaches. The human element looks at far more than the designer or human operator. It includes supervisors, managers, staff associates, personnel directors, purchasing departments, etc. It could be the software programmer who makes a single digit mistake out of tens of thousands, when massaging a complex piece of software.

What we face in operator–process–product–interface is the human ability to meet system requirements. Figure 1 illustrates our status. When we design a process it makes various demands (dashed line) to make it work effectively. The operator (solid line) does not perform on a constant, straight line basis. When we place the lines on top of each other, where they intersect is a mishap, an undesired, unwanted event. On top of that are the system limits, the parallel straight lines. System or human limits can also form a mishap point by exceeding the process or system parameters. The circles show mishap points.

Now, we can exchange nearly anything for the human operator in this diagram, a training program, design data, supervision, management, government resources, etc. Whatever we substitute, the effect is the same, an undesired, possibly catastrophic event.

UNPLANNED INTERFACES

We have complex problems that must address complex systems, on the forward edge of our knowledge. An example is separate, often unrelated systems that can impinge on each other but, cannot be forecast.

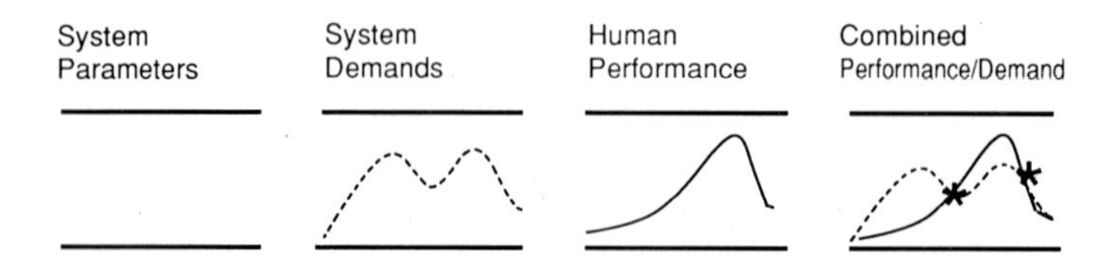

Fig. 1. Performance and demand with system parameters.

New Zealand Airways

The New Zealand Airways mishap, a few years ago, shows how the software–human sub-system interacts to cause a mishap even after all engineering tests, etc. The mistake was a single keystroke in writing a software program for a self-contained navigation system. The system, after months of actual use, sent a large transport aircraft into the side of an Antarctica mountain, killing all on board. As usual, many other human errors were present in terms of management, staff, engineering checks, lax government control and pilot discipline. The mishap nearly brought down the New Zealand government.

Interstate Bank

The Interstate Bank high-rise building in Los Angeles caught fire ten days before a new sprinkler system became operational. A very well designed fire-fighting system was not yet working. Engineers did a good design and installation job. Whose fault is it that it was not yet in working order? This was an unplanned interface. Fortunately, the fire was in late evening when few tenants were on the job. Down in the lobby the security force was monitoring sensors. They received warning of a problem on the 10th floor of the 15 year old building. Based on prior false warnings it was dismissed as a system fault. A second warning, shortly after, was also disregarded. On the third warning a man took an elevator up to check things. He confirmed a fire and the fire department reached the scene within three minutes. But for a heroic, life threatening effort by the fire department the entire building would have been lost. As it was, the man in the elevator was the only fatality. The building and hundreds of businesses were out of action for weeks. Fortunately, the building was well engineered and repairable. If the first warning had been checked the fire would have been confined to one or two rooms and a life saved.

Space Shuttle

The U.S. lost their space shuttle in a fiery explosion three years ago. The trouble was traced to an "O" ring not designed to handle freezing temperatures. A thorough investigation detailed the lack of certain engineering precautions and analysis. Still, management and staff decisions permitted the catastrophe to occur. Even now, the management and staff contributions are not fully exposed. We did chastise some designers and engineering managers at the site of rocket design and manufacture. But what about the staff and management decisions that encouraged, yes – permitted the risk to be taken?

Chernobyl

The Chernobyl nuclear accident began when "...the reactor's (power) suddenly increased during a scheduled shut down ... The considerable emission of steam and later reaction resulted in the formation of hydrogen, its explosion, damage to the reactor and the associated radioactive release."

The report said the disaster was mainly the result of a breach of safety regulations. Ironically, the mishap began the day before with an experiment designed to test the safety of the plant during a total power failure. Together, the plant operators made six major mistakes. "If one *violation* out of six had been removed, the accident would not have happened."

DOUBT IS CAST ON OUR PROCEDURES AND COMPUTERS

No shadow is cast on hundreds of thousands of highly trained and capable engineers who design systems and their components. They do their best. But, who looks at the complete systems and the interaction between them. These are staff, management, and technical expert matters to coordinate.

The many implications of computer integrated manufacturing (CIM) for human workers should be considered at the earliest planning stages to assure peak performance from a totally integrated system. Systems' designers are often too preoccupied with robotic and other high-tech hardware and software that the human component is forgotten. This mistake dooms an otherwise well conceived system to failure as workers struggle to operate safely and maintain systems inadvertently designed to thwart their efforts.

Human aspects in design and use of CIM systems fall into three areas where impact may be felt. They are installation/implementation, operation/maintenance and safety. These areas should first be considered in the conceptual or initial design stage, not later.

The inherent hazards in traditional manufacturing, material handling, equipment and computers are addressed in the literature. The well-known principles of successful safety programs must be extended to cover cases which arise with new manufacturing scenarios.

The injury potential, and thus the need for attention to human aspects in design, exists wherever human workers and equipment interact. This occurs mainly during routine operation, programming, maintenance and through accidental contact. Accidental contact may be due mainly to equipment failure, the program or the human worker.

Computers can help, but ...

We commonly overlook, in computer use, that a central goal of design is to anticipate failure. Thus, it is vital to know exactly *how* a structure may fail. The computer cannot do this by itself. There are attempts to place artificial intelligence into the machine to make it an "expert" system. Some dream that the ultimate computer assisted design (CAD) will have the computer learn from experience kept in a computer file of failures. However, when such a notion becomes reality, the engineer must still ask crucial questions. Will this improperly welded pipe break if an earthquake hits the plant? Will the auto body crumple in this manner when it strikes a wall at ten miles per hour? One can question the computer, but whether the questions are asked depends on the same human judgment that dismisses the fatigue question in many mishaps.

While the computer works quickly as a file clerk, it cannot work quickly when asked certain engineering questions. One of the most important questions in design concerns metal behavior under loads that deform components permanently. It takes only seconds to put a bar of ductile steel in a testing machine and pull the bar until it stretches and breaks. Simulating such a basic test on the largest computer takes hours and hours. The human engineer must still make a judgment, just as in the old days, as to which way a material is likely to fail. Our inability to use our computers fully may stem from our reluctance to exchange information. You can see it in the *Proceedings of the First International Conference on Computing in Civil Engineering*. One session was given to "Anatomies of Computer Disasters". This abstract covers the session:

> "No papers for this session will be published. The purpose of this is to permit the speakers to be very candid regarding the various computer disasters which they are describing. Names, organizations, etc. will not be used in order to protect the privacy of those concerned."

This is not professional of the engineering, safety, legal or any other profession.

Engineers design individual elements or systems of hardware to meet human needs and wants. Engineering practice draws on principles in the physical sciences, seasoned by empirical experience. It is then subject to professional judgment in tradeoffs to reconcile conflicting requirements and constraints. On its own, the engineering profession advanced from an applied science to recognizing that its clientele are people. It was seen that graduates needed exposure to the humanities and liberal arts. Engineering curricula were then broadened. A more recent external force has been public expectations of the profession to protect human health and safety. While the profession asks if they should indeed do this, new challenges arise from science advances that translate into hardware. Many of these newer challenges come from software and "socialware". That translates to organizations and management instructions crucial to effective working of the hardware.

The twin disasters of Challenger and Chernobyl make the point that although failures are attributed to hardware, the most critical lapses may be in "socialware". The point is that these varied functions must be laced with communications networks designed to work to some socially optimum criteria.

The engineering profession must be necessity deal with much more than technology. They must also deal with the human system. Dealing with the "whole thing" requires attention to the dynamics of a system represented by the question, "Can we manage it?". The answer requires more than a technological fix. Preparation to do this will likely call for a revised educational system for engineers.

NEED FOR AN IMPROVED OVERVIEW SYSTEM

We need a way to collaborate, integrate, overview, bring together for consideration, and combine all aspects of a system and possible interacting systems – even if they do not seem related. We need a way to consider all aspects of the engineering, social, political, governmental and corporate environments. There must be a way to influence, and be properly influenced, by lesser and more knowledgeable managers and staff. We need ways to enter the latest knowledge into our planning and decisions and know we are on the cutting edge of knowledge. This almost impossible task increasingly faces us and is impossible to avoid. We need an overview system of checks and balances that studies every interface and ensures that all risk elements are considered. We need persons who can see the big picture, overview, coordinate, assimilate, and bring every aspect of risk into focus.

We know that in any mishap that a management element is always present. Let us suppose you are designing a bridge, likely under contract to a local or state government. By sheer economics we have constraints on the hours spent making the design safe. There is only so much money for so many hours of work. When asked to do the impossible, that is without enough money, time, or experience – that is the fault of the management and staff system. How does that affect our bridge design? In any one of our countries somewhere between 60 and 80 percent of our bridges are deemed unsafe by modern standards. If this is not our fault, whose then is it? Ask that question and we bring in a whole range of planners, designers, financiers, managers, staff consultants, etc.

Hyatt Regency Hotel

The problem is seen in the walkway collapse at the Hyatt Regency Hotel in Kansas City during a 1981 tea dance. It killed 114 people and injured 200 others. Design changes and lack of inspection made the disaster inevitable. The walkways that collapsed could barely support their own weight, due to a modification made during construction. The disaster review board stated: "Unfortunately, the demands of our profession are such that no one person can check all the work". That from a man who had helped design the Hyatt skywalk across the ballroom.

Kemper Arena

Still later in Kansas City, a large domed arena collapsed when high strength steel bolts failed under load. The particular bolt has plenty of strength – except it loses it's strength when it is loosened or tightened a few times. Must we communicate this word, on that type bolt (A-490), over and over? Is this an engineering, management, supervision, communication, human factors, or what kind of problem?

A new building is a one of a kind affair, a custom job. There are dozens of specialists and hundreds of other people involved. There are millions of connections to be bolted, welded and sealed. This has brought a new expert on the scene, a diagnostic engineer. It is his job to find things wrong with new structures, after they are built and before they are in use. His tools are common sense, deductive skill and technology. Unfortunately, he is not superman enough to understand all the human interactions that can happen after the product or process is in place and in use.

King's Cross

The report on London's King's Cross Underground Tube fire disaster that took 31 lives severely criticizes management. The 91 day hearing found 40 examples of shortcomings before and during the accident. Safety systems are criticized as reactive and unconstructive. Said the counsel to the hearing: "There was no positive seeking out of enhancement of safety from and in fires". He further said that: "Standards had been allowed to slip. Standards from the highest to the lowest levels in London Underground had been allowed to slip ... Human defenses were not satisfactorily in place." While the staff were imbued with a safety philosophy, an active approach to safety was not in place. It was reactive. What was lacking, the report said: "... was the correct perception and direct assessment of risks ... The risk of fire was also not adequately evaluated". The Counsel went on to note: "In particular, the King's Cross disaster illustrates an urgent need for significant change in the thinking of most senior management levels that safety integrity is fundamental and a precondition of operating the system". The results of these should have been seen along with the possibility of a fire. While engineering solutions could have eased some of the problems, most aspects had supervisory and management roots.

Space capsule

I use many American examples to make my point. Our tragedies and catastrophes, and our many errors qualifies their use. Forgive me if I do not use your well-known catastrophes. The best documented example of our early failure to overview all aspects of a situation concerns our first major space tragedy. In the 1960s, many people had urged a safety overview system for our space efforts. The NASA engineering faction, working under a "rush" psychology, strongly resisted the program. They argued that they had doubly considered everything using the finest engineering techniques and built in redundancy. Only a week before the 1967 tragedy, NASA, for the umpteenth time, refused to integrate a safety overview system.

Then, when the launch pad mishap took the lives of three astronauts, public uproar forced an in-depth investigation that found many, management and engineering oversights. The mishap caused NASA to activate a safety program, even before the investigation was well underway. Under the guidance of Jerome Lederer, perhaps the most detailed safety oversight program ever seen was placed in action. Most, but not all, of their recommendations were followed. The program mostly emphasized, safety communication systems, a new safety survey system, use of system safety techniques and safety awareness programs. The U.S. then went on to 20 years of unparalleled space successes before the oversight effort deteriorated and old problems resurfaced.

Bhopal

Chemical processes introduce new considerations, but the same human frailties underlie the engineering processes. The worst industrial disaster in world history, Bhopal, happened because an industrial design tolerated a very large inventory of a highly lethal chemical. It appears that water found its way into one of three storage tanks and an exothermic reaction developed. A tank refrigeration system had been shut off for six months and there was nothing to slow down runaway reactions.

Let us assume that there was nothing wrong with the design of the Bhopal plant or process. The catastrophe still could have happened because of poor communication, disregard of proper processes, lack of proper training, and inadequate supervision and management. The affair could have been prevented or ameliorated if one of five backup processes were in working order; a high temperature alarm, a vent gas scrubber unit, the burning tower, pressure valves, or a tank refrigeration unit. Their operation are not normally engineering problems, but they must be part of risk assessment. They are gaps and vacuums in the design to disposal process.

"It is difficult to conceive of any failure or difficulty – major or minor – that does not fit into one of these classifications ...":

1. *Ignorance* – incompetent design, construction or inspection.
2. *Economy* – supervision and maintenance by persons without necessary intelligence.
3. *Lapses*, or carelessness – an otherwise competent designer/architect shows negligence in some part of his or her work.
4. *Unusual* occurrences – earthquakes, extreme storms, fires, etc.

In 1982 a U.S. government committee on Science and Technology examined structural failure problems in the U.S. Their findings covered many things, but they listed six critical features in *preventing* structural failures.

1. *Communication and organization* in the construction industry.
2. *Inspection* by the structural engineer.
3. General *quality* of design.
4. Failure to *connect* structure with design details and shop drawings.
5. *Selection* of architects and engineers.
6. Timely *circulating* of technical data.

In bringing together all things to consider in a system, I find little better than the Management Oversight and Risk Tree (MORT) System developed and used by the U.S. Department of Energy. Developed mainly to reduce the risk of nuclear mishaps by an order of magnitude it applys to any type of operating system. MORT is not the final answer, but it includes the most complete system of checks and balances we have devised. Do not be misled by the emphasis on the word *management* in the approach. MORT can be as highly technical as you wish, when worked in depth.

We seldom reach perfect safety because to the limits on our knowledge and ability. There are still ways to oversee a new process or design that ensure the most safety under the circumstances. Let me challenge you with some simple examples.

GAPS AND VACUUMS IN DESIGN TO DISPOSAL PROCESS

Figure 2 shows a typical look at our product or process life and the areas where engineering approaches solve problems. This leaves many gaps and vacuums in knowledge (Fig. 3). Throughout the life of our product or system there are areas, here shown as question marks, we do not approach for lack of understanding or resources. These question marks are gaps and vacuums in our processes.

OVERSIGHT APPROACH AND CONTROL

We need administrators/managers for this task, "overseers" of risk assessment and risk prevention. We need persons with technical insight and oversight capability who can see the big picture and the myriad of issues facing risk prevention. These overseers will apply the control function. In business and industry this includes all activities of a manager to assure that actual results meet planned results. Controlling covers all company operations and all man-

Design ⇨ Prototype ⇨ Test ⇨ Operation ⇨ Life ⇨ Disposal

Fig. 2. The life cycle.

? Design ? ⇨ ? Prototype ? ⇨ ?? Test ? ⇨ ?? Operation ? ⇨ ?? Life ? ⇨ ?? Disposal ??

Fig. 3. Gaps in the life cycle process.

agers such as chief executives, financial managers, personnel managers, or engineering managers. In seeking quality from concept and design to disposal we include certain control factors and a management system.

Model process

Suppose we design a new production system, either simple or complex. It is not only a matter for process and design engineers. Someone must also look at specific control factors.

Specific Control Factors

Technical information
Facility functions
Maintenance
Inspections
Supervision
Middle management
Staff support

Every item above can be examined in detail. If you think that this is too simplistic, let us briefly look at one thing on the list, the items to consider under the "technical information" system.

Technical information

The following is an example of the things to consider under the heading of "technical information". Any of the other factors, facilities, maintenance, etc., can be examined in the same depth.

Knowledge
Known precedents
Internal communications
External communications
Monitoring systems
Safety observation plan
Accident/incident system
Significant observation system
Error sampling system
Safety inspections
Upstream process audit

Management system

Not one of those "control factors" operates in a vacuum. They are driven by and related to a management system. Thus, any breakdown in the control fac-

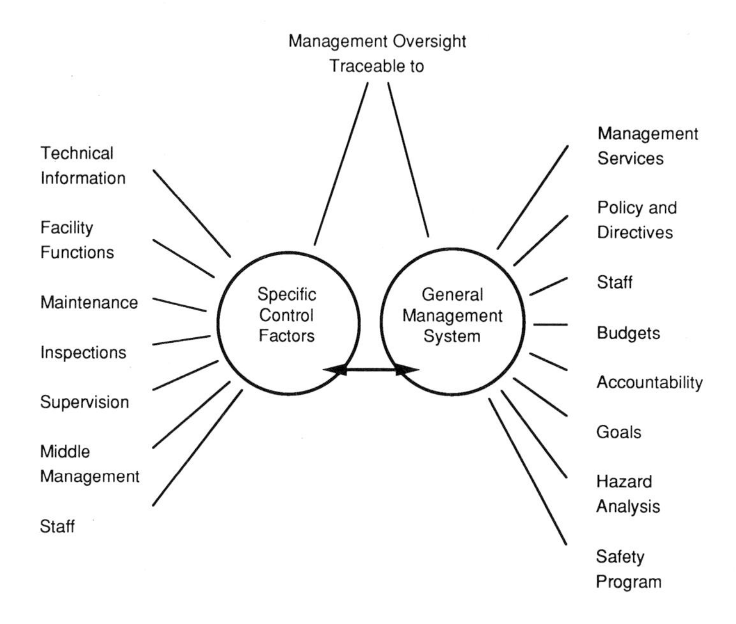

Fig. 4. Management oversight relationships.

tors relate to failures in a management system. Related management system factors include:

Management services	Accountability
Policy and directives	Goals
Staff	Hazard analysis
Budgets	Safety program

√ *Management services*

To illustrate some of the factors in the management section that relate to our technical control factors let's examine the first, "management services". What specifically might we look at here?

Research and fact finding	Training
Information exchange	Technical assistance
Standards and directives	Program aids
Deployment of resources	Measure of performance
Risk response	Coordination
Resources	

Our "oversight" approach to risk prevention then considers both the control factors and the management system. Since any problem with control factors can be traced to management factors (see Fig. 4).

It goes back further than this when management assesses the risk prevention program. Based on their assessment they either correct faults and problems or ASSUME THE RISKS that go with it. Perhaps management decides that resources invested in marketing will provide a better return than more resources in technical expertise or an extra person to do fault tree analysis. They then assume the risks of not providing your resources.

I have just given you excerpts from an already well-known and proven overview system called Management Oversight and Risk Tree, MORT for short. It is not the only system or a perfect system, but it does look at the complete operating system and interacting systems. There is another way of looking at the system in operation, in our example, a process system.

MAIN ELEMENTS OF JOB PROCESS CONTROL

The well-known elements of job process and procedure control are seen in Fig. 5. We can design the finest technical procedure or hardware, but unless we consider the people, it will not work as designed. Unfortunately, many of these three areas are beyond our control. Let me show what I mean. Look at people considerations:

People
- Selection
- Training
- Testing and Qualification
- Evaluation

Each of these items is nearly a discipline in itself. It is impossible for our group to consider them in detail, unless that is our speciality field. Consider the hardware part of the triangle.

Hardware
- Concept and design
- Life cycle study
- Fabrication
- Installation
- Occupancy-Use
- Operation
- Modification
- Decommission and disposal

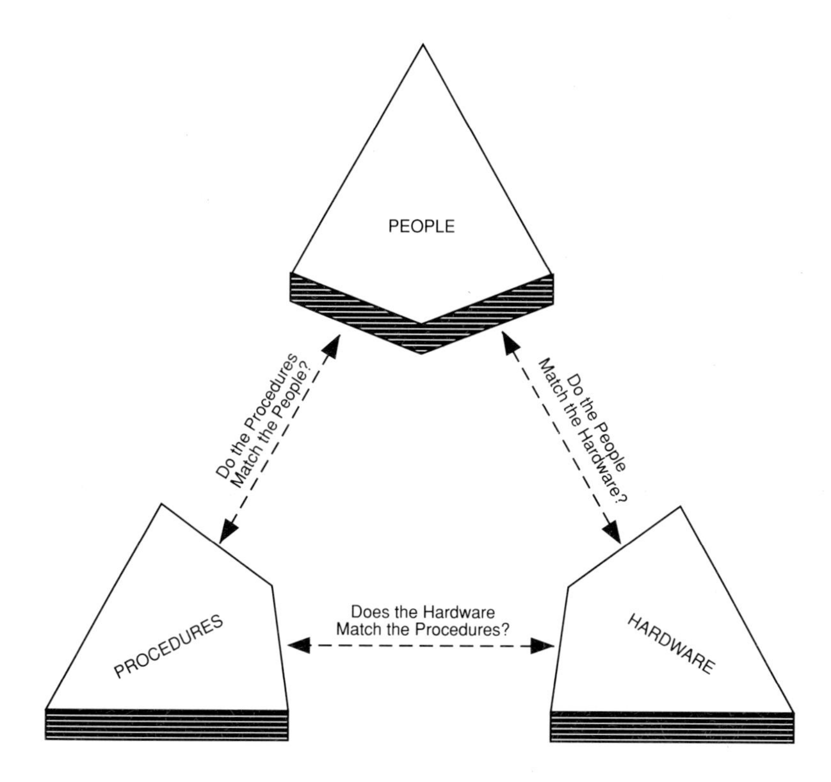

Fig. 5. The main elements of job control.

Procedures

Our conference directly addresses people, hardware and procedures. But, what about the details of procedures as we have just shown for people and hardware? More important, what about the interfaces between them? Let us look at "procedures" and see. Effective procedures have three aspects we often overlook in their development; participation, testing and feedback.

Participation. Risk preventing procedures cannot happen in a vacuum, or even on the drawing board. While engineers, designers and managers may write the safety procedures, they are not the ones who carry them out. The worker is the one who does it, but her/his participation in our development of procedures is slight.

Testing. Procedures developed and committed to paper must be validated. They should be tested in an actual workplace by workers who will use them.

Feedback. Change is a constant. It is always taking place. Your procedures require periodic adjustment. This requires an good feedback system that goes from top management to the worker and back again. Are you, the engineer, in the feedback loop?

OVERSIGHT ROLES IN DESIGN THROUGH DISPOSAL

The complexity of a complete overview system is seen in a study by the Systems Safety Development Center at the Idaho Nuclear Laboratory in the U.S. They are now processing a white paper, *Environmental Performance Indicators and Root Cause Analysis*. Their rationale states that the risk analysis tools of "performance indicators and root cause analysis are related to the larger picture of risk management". I quote: "The priority need in Environment, Safety and Health risk management is to gather together bits of risk analysis method and technology into a risk management 'system'. This involves integrative development work in a number of areas." A determined effort is underway to build the complete oversight program we need for risk prevention. I have great faith in the Idaho group because of their landmark actions to improve safety. Their open-mindedness goes beyond national boundaries and they are committed to handling complex, interacting systems with a program having wide and practical use.

I am not selling MORT or any overview system. My purpose is to emphasize that to succeed in risk prevention, we must consider all elements of a system, and the interacting systems. This requires an approach that views the larger system – considers everything in it. That demands a special manager who can see and understand the complexity of problems associated with total risk assessment and prevention.

CONCLUSION

We find many gaps in our knowledge and capability for risk prevention and in the design through disposal concept. The gaps range from human interactions to social and political influences. Risk assessment by comprehensive engineering from the early phases of the product design through disposal is only part of the answer. The interacting elements of the solution are so complex that it defies an approach by one or a few disciplines. The problems cannot be ignored no matter the difficulty of addressing them. The solutions are within our expanded overview capability.

We need an overview system, an umbrella, that considers all aspects of technical, managerial, and support staff. This in turn calls for a very special person of broad capability to tie such an overview together. That overview system will certainly be as broad as the Management Oversight and Risk Tree, perhaps more encompassing.

There is no easy solution to our problems. Comprehensive engineering practice is not enough to cover human and management interactions. I challenge you to keep the overall problem in mind and not be handicapped by traditional thinking and approaches. If I do not bring you solutions, only more problems, there is no apology. It is a complex mission. I do suggest an interdisciplinary overview system that addresses the complete task. My assignment was to bring the gaps and vacuums in our programs and processes to the forefront – for your action.

FURTHER READING

Anon., 1988. Autumn Verdict on King's Cross. Occup. Saf. Health, London, August: 000–000.

Union Carbide Corporation, 1985. Bhopal methyl isocyanate incident investigation. Union Carbide Corporation, Danbury, CT.

Brandsje, K., 1988. Being there. Fire J., 82(5): 66–71.

Brown, C.J., 1981. Proc. 1st International Conference on Computing in Civil Engineering, New York, American Society of Civil Engineers.

Donnelly, J.H., Jr., Gibson, J.L. and Ivanchevich, J.M., 1987. Managing work and organization. In: Fundamentals of Management. Business Publications, Inc., Plano, Texas, New York, pp. 208–225.

Ferry, T., 1988a. Modern Accident Investigation and Analysis. Wiley, New York, 2nd edn.

Ferry, T., 1988b. Safety Management Planning Manual. Merritt Company, Santa Monica, pp. 2–9, 8–1.

Gorbachev, M., 1986. Televised speech in Moscow on May 14.

Hall, S.K., 1988. Industrial chemical disasters. Prof. Saf., Des Plaines, IL, July, pp. 9–13.

Jarvis, S., 1988. London underground's safety management philosophy scrutinised. Saf. Manager's Newslett., London, July/August.

Larson, E., 1988. A new science can diagnose sick buildings before they collapse. Smithsonian, May: 116–127.

Lederer, J., 1988. Personal correspondence, August 5.

Legasov, V., 1986. Soviet report made to the 1986 International Atomic Energy Agency (IAEA) Conference, Vienna, Austria, August.

Nertney, R.J. and Briscoe, G.J., 1988. Draft ES&H Performance Indicators and Root Cause Analysis - A White Paper. System Safety Development Center, Idaho Falls, ID, September 1.

Petroski, H., 1985. To Engineer is Human; The Role of Failure in Successful Design. St. Martin's Press, New York, pp. 53, 81, 86, 109, 112, 115, 195, 205, 207, 208, 210, 211, 219, 224.

Rummel, P. and Holland, Th.E., Jr., 1988. Human factors are crucial components of CIM system success. Ind. Eng., 20(4): 36–42.

Wenk, E., 1988. Portents for reform in engineering curricula. Eng. Educ., 78(10): 99–102.

How people learn to live with risk: prediction and control*

A.R. Hale

Safety Science Group, Delft University of Technology, P.O. Box 5050, 2600 GB Delft, The Netherlands

ABSTRACT

Hale, A.R., 1990. How people learn to live with risk: prediction and control. *Journal of Occupational Accidents*, 13: 33–45.

Models are presented of the deviation process which leads to failure and accidents, and of the way in which human action intervenes to provoke and/or recover from deviations. The three levels of human information processing and their typical error types are summarised. The models are used to derive lessons for system designers and system managers in producing and running systems which will be robust against failure throughout their life cycle.

INTRODUCTION: HUMANS IN SYSTEMS

In order to understand the role of the human factor in risk control it is necessary to have a clear model of the area to be talked about. The theme of this conference is a total approach to risk control based upon consideration of the total life cycle of the system under consideration: from design to disposal. In discussing the human factors in this cycle it is convenient to split it into two parts:

the conceptual phases:
- original design
- later modification

the active phases:
- construction/installation
- use
- maintenance/cleaning/adjusting
- disposal/refurbishing

---

*Presented at the International Conference on Industrial Risk Management, Zürich, Switzerland, 16–17 January 1989.

The first part occurs in the head of the designer in interaction with the drawing board or CAD system and dangers are things to be conceived of, predicted and designed out. The designer is looking after the safety of others as part of a demanding intellectual and creative exercise.

The second part occurs in reality and dangers are controlled or let loose through the decisions of operators, managers and others about their own and other people's safety.

The designer in the first phases tries to predict and control the decisions and actions in the second. Figure 1 gives an indication of the steps that are important (Hale and Glendon, 1987). The designer tries to build into the choice of technology, the choice of materials and the detailed design of the system as few inherent dangers as possible (as little energy or toxic material which can escape).

Where it is not possible to eliminate danger sources totally from the system the next line of defence is to build in control functions. These are either passive barriers to the escape of the energy (e.g., thick walled pressure vessels) or active controls which monitor the normal functioning and intervene when it moves outside defined parameters (e.g., thermostats, pressure regulators).

At a certain moment the system will move outside those defined design parameters in such a way that the in-built controls can no longer recover the

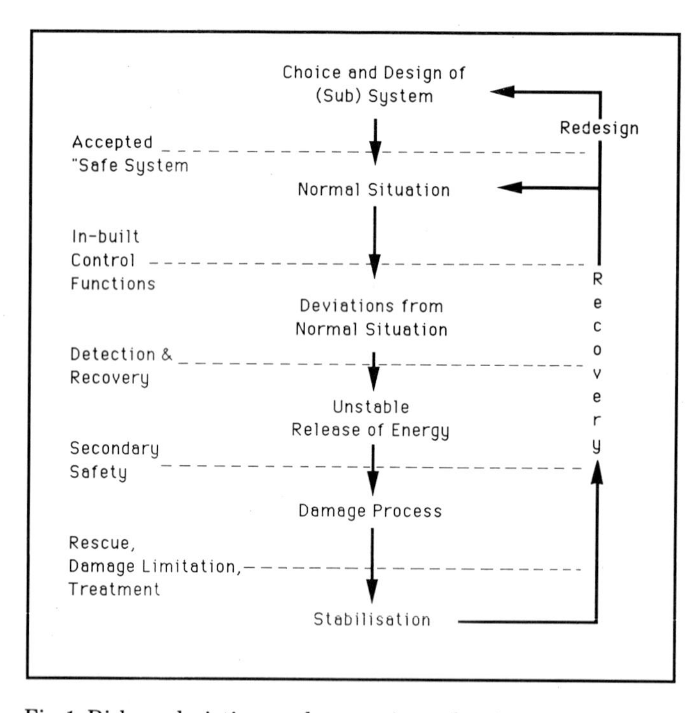

Fig. 1. Risks as deviations and prevention as barriers.

situation. In complex high technology systems with sophisticated control equipment such a situation will occur quite rarely; in most other systems it occurs frequently, if not constantly (compare a train with a car in this respect). The deviation may be caused by:
- factors not considered by the designer
- failure of one or more hardware elements
- a deviation from required operating rules by a person.

The system moves now into a metastable state which may last anything from seconds to months in which some factors, but not all, are present for the occurrence of harm. In this period of time there is opportunity for detection of the hazard and recovery, almost always by the human elements in the system (the operator, maintenance crews, inspectors, etc.). If recovery does not occur the system will eventually move into an unstable state, when the last determinant for the occurrence of damage occurs (the final straw which breaks the camel's back), and only secondary safety measures to contain or divert the energy which is released can be of use in limiting damage to the less important system elements and letting the more important escape.

With this model as background it is possible to define the role of humans in the process of risk control (see Fig. 2) as:

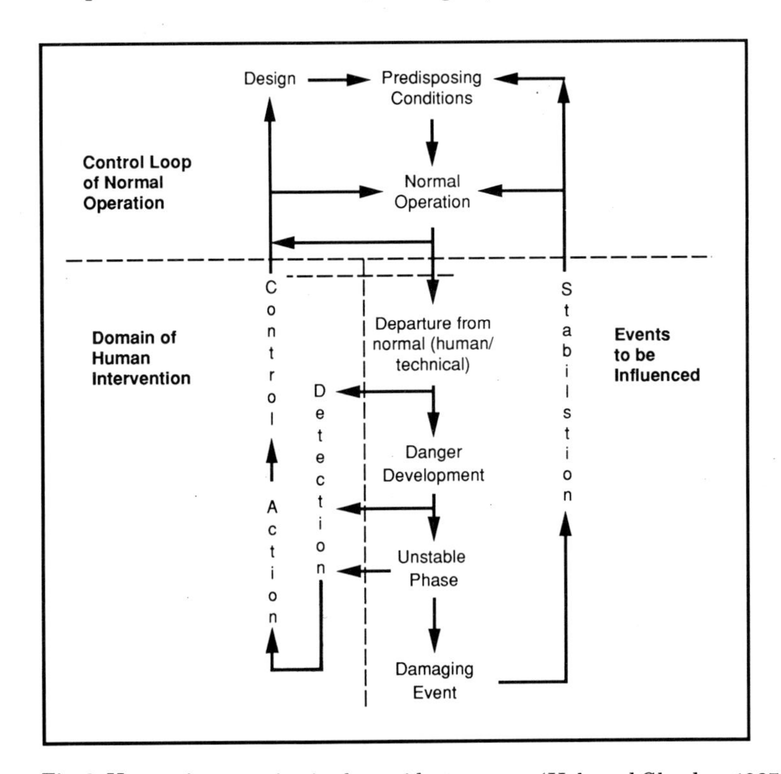

Fig. 2. Human intervention in the accident process (Hale and Glendon, 1987).

- predicting and designing out dangers
- (avoiding) precipitating deviations from normal operation
- detecting and recovering from deviations which do occur
- running away as fast as possible from escaped energy.

It is now necessary to look at the way in which people are able to respond to these tasks.

MODELS OF HUMAN BEHAVIOUR

The work of Reason (1987) and of Rasmussen (1980) has added a new dimension to the modelling of human behaviour in the face of danger. When combined with older models of the individual as a processor of information (Hale and Perusse, 1977) we can produce much more sophisticated models of how people react to the task of predicting and controlling danger. Figure 3 is such a model set out in the form of the questions which have to be answered positively at all times if individuals are to remain in control of the dangers in their environment. Since there is no such thing as a risk-free environment, this is a constant part of our activity as human beings.

It is convenient to distinguish three levels of behaviour which show an increasing level of conscious control:

1. *Skill-based* behaviour in which people carry out routines on "automatic pilot" with built-in check loops.
2. *Rule-based* behaviour in which people select the routines, at a more or less conscious level, out of a very large inventory of possible routines built up over many years of experience.
3. *Knowledge-based* behaviour where people have to cope with situations which are for them novel and for which they have no routines. This is a fully conscious process of interaction with the situation to solve a problem.

As a working principle people try to delegate control of behaviour to the most routine level at any given time. The direction of the arrows in the figure therefore indicates the sequence with which we operate. Only when we pick up signals that the more routine level is not coping do we switch over to the next level. This provides an efficient use of the limited resources of attention which we have at our disposal, and allows us to operate as parallel processors. We have, hereby, a certain capacity for progressing two jobs at once, provided that they do not both demand the use of the limited conscious capacity at the same time. We can concentrate our attention on those aspects of the task in hand which interest us, or where we are currently having problems, and let the rest of our activity more or less look after itself. The crucial feature in achieving error-free operation is in ensuring that the right level of operation is used at the right time, whereby it must be stressed that it can be just as disastrous to operate at too high a level of conscious control as at too routine a level.

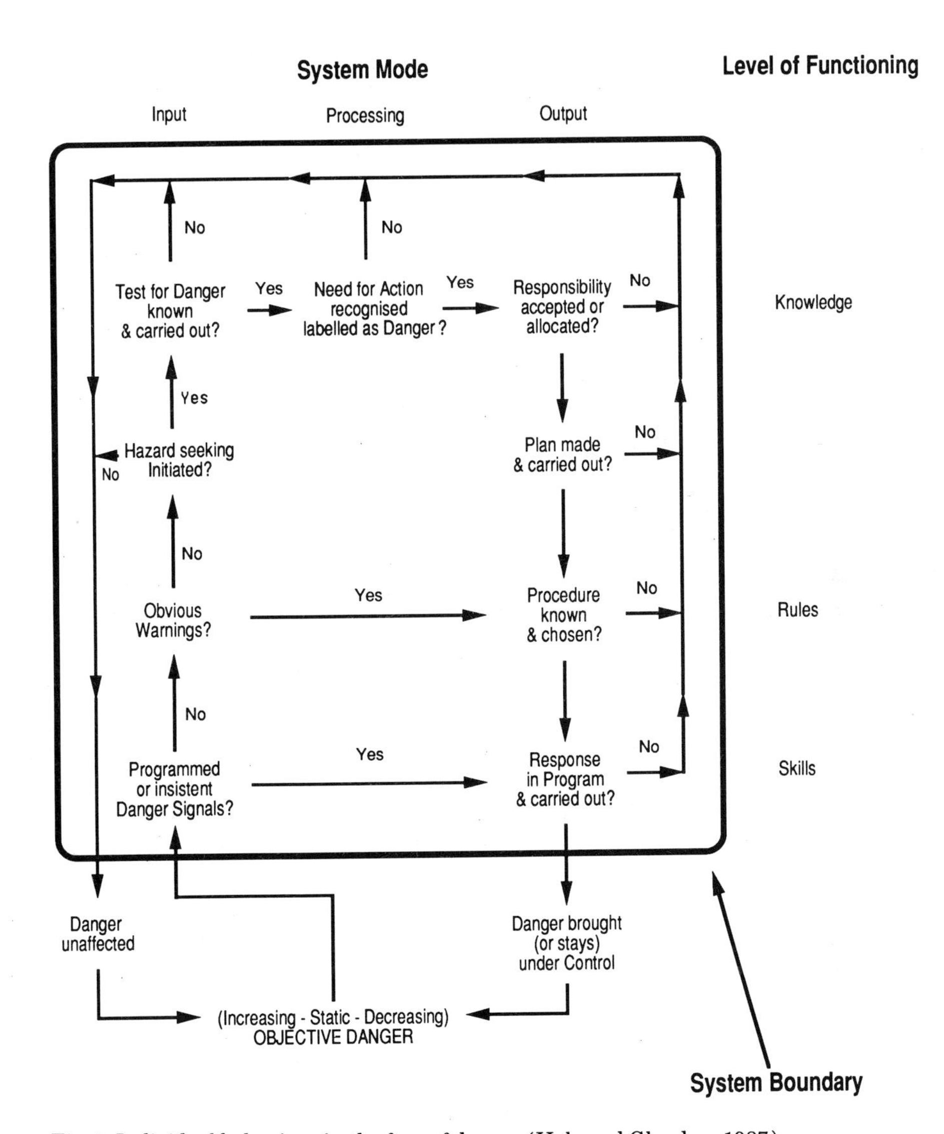

Fig. 3. Individual behaviour in the face of danger (Hale and Glendon, 1987).

Each level of functioning has its own characteristic error types, which are described briefly in the following sections.

1. Skills and routines

All routines consist of a number of steps which have been highly practiced and slotted together into a smooth chain, where completion of one step automatically triggers the next. Built into the chain are test loops which check that the expected result of an action has occurred. The routine dangers which are constantly or frequently present in any situation are (or should be) kept under control by building the necessary checks and controls into the routines as they are learned. Examples are the skills of laboratory technicians conducting routine tests with hazardous materials, fitters dismantling machines, or electricians working on live apparatus. The checks still require a certain amount of attention and the comparatively small number of errors which occur typically at this level of functioning are ones where that attention is disturbed in some way.

1. If two routines have identical steps for part of their sequence, it is possible to slip from one to the other without noticing. This nearly always occurs from the less frequently to the more frequently used routine; for example arriving at your normal workplace rather than turning off at a particular point to go to an early meeting in another building. Almost always these slips occur when the person is busy thinking about other things (e.g., making plans, worrying about something, under stress) and is not alert enough to switch out of the more familiar routine at the right moment. This planning activity can itself result in a shift from one routine to another in some cases, where the person is consciously planning the next step in the actions and begins on it before completing the current step.

2. If someone is interrupted half way through a routine they may return to the routine at the wrong step and miss out a step (e.g., a routine check) or carry out an action twice (e.g., switching off the instrument they have just switched on because both actions involve pushing in the same button).

3. Another version of this type of error is what Reason calls the "nosy supervisor". If you try to check consciously how a routine is going you can disturb its rhythm, because the process of conscious checking has a much slower dynamic which disrupts the smooth, fast skill. Anyone who has tried to teach their child the skill of a sport or musical instrument at which they themselves are proficient knows this problem.

4. The final problem at this level is that routines are dynamic chains of behaviour and not static ones. There is a constant tendency to streamline them and to drop steps which appear unnecessary. The most vulnerable steps are the routine checks for very infrequent problems in very reliable systems (e.g., tapping the face of a modern measuring instrument to check that the needle has not stuck).

It should be clear that many of these errors occur because the boundary between skill-based and rule-based activity has not been correctly respected.

These sorts of error will be immediately obvious in many cases because the next step in the routine will not be possible; the danger comes when the routine can proceed apparently with no problem and things only go wrong much later. In that case the error will be very mysterious to the person who made it, because they will think back (at a conscious level) over their actions, they know very well how the actions should be carried out (they are by definition well practiced routines), and they will also (again by definition) not have been conscious of the action sequence actually carried out. However, the cure for the errors does not lie in trying to make people carry out their routines with more conscious attention. This will precipitate more "nosy supervisor" errors, will take too long and so be too inefficient, and will be subject over a short time to the erosion of steps mentioned above.

The hazards which cause the most problems are ones which do not occur frequently (and so the steps to control them become eroded) and ones where the consequences of incorrect action are not apparent for a long time (e.g., the effects of bad manual handling methods and postures on muscles and joints, the effects of noise on hearing, the effects of handling (or smoking) chemicals on lungs or as carcinogens). In all of these cases the steps to control the hazard are either not learned or are too rapidly forgotten unless frequently practiced.

The cure for the problems lies to a great extent with the designer of the routines (and so of the apparatus or system) to ensure that routines important for safety (e.g., emergency shut-down routines) are either identical with highly practiced routine, or very different so that unintended slipping from one to the other is avoided. Where such confusion is unavoidable it is often possible to build in extra signals to warn that the wrong path has been entered by mistake. Such signals can be ones from automatic instruments, or ones from another member of the team working on a problem, who has the specific task of monitoring critical points of this sort. This makes use of the normal in-built checking routines which are in most cases already very efficient. This approach gives a clear task for the system designer to predict these errors at the (knowledge-based) design stage.

The second line of defence is to train people thoroughly so that the correct steps are built into the system, and then to organize supervision and monitoring (by the people themselves, their work or reference group, and supervisors or safety staff) so that the steps do not erode.

2. *Rules and diagnosis*

When the routine checks indicate that all is not well, or when there is a clear bifurcation needing a choice between two (or more) possible routines, people must switch to the rule level. Choice of a routine implies categorisation of the situation as "A" or "B" and choice of routine X which belongs to A or Y which belongs to B. This is a process of pattern recognition.

The errors which people make at this level are linked to the way in which people normally function as opposed to the way that scientific principles say they should function. We all have the tendency to formulate hypotheses about the situation which faces us on the basis of what has happened most often before (I think that the alarm bell is ringing because they are testing the fire bells) and then seek evidence to confirm that diagnosis (it is 12.00 on the first Monday of the month when they always do it). This is in contrast with the scientific method which bids us seek evidence that would not be compatible with the hypothesis (is there any sign of smoke or other people evacuating buildings?). This means that there is again a conservative tendency in actions; people tend to think they are facing well-known problems until they get unequivocal evidence to the contrary. The Three Mile Island was a classic case of this where operators persisted with a false diagnosis for several hours in the face of contradictory evidence (which they explained in another way), until a person coming on shift (and so without the mental block of having made the initial diagnosis) detected the incompatibility between symptoms and diagnosis.

The development of decision support systems in the process industry tries to combat this tendency by getting operators to build expert systems of the ideal "production rules" (e.g., IF A, B, C and D occur, THEN the problem is probably P, and that can be checked by observing Z). This construction activity can occur in periods where the operators have plenty of time. When the emergency occurs the operator can then feed in the preliminary diagnosis, which the expert system checks against its rules and the indications from instruments to detect if all the symptoms present are compatible with the hypothesis. If not it informs the operators and asks them to think again. Note that it does not take over at this point like a fussy teacher and tell the operator he or she is wrong. That would precipitate the reaction: "if the system is so clever let it cope; why should I bother to do anything". That is an attitude which will lead to more disasters when the expert system gets it wrong too. A simple example of a decision support system which adopts the right approach in this is the speller check delivered with your word processing package.

Again the errors come because the boundary between two modes of operation is not properly respected. In this case people remain at too routine a level instead of consciously accepting that the situation facing them is unusual. The solutions lie in giving alarm signals to help them to switch somewhat earlier.

3. Knowledge and problem solving

When people do switch to the fully interactive problem solving stage they have to rely upon their background knowledge of the system and the principles on which it works to derive a new rule to cope with the new situation. There are meta-rules for problem solving which can be taught. University education should be about learning such meta-rules. Besides these there is the creativity and intelligence of the individual and the thoroughness of their understanding of the principles underlying the machine or system (technical, organisational or political) they are trying to influence. Errors at this level can be traced to:

1. Inadequate understanding of these principles (inadequate mental models).

2. Inadequate time to explore the problem thoroughly enough.
3. The tendency to shift back to rule-based operation too soon and to be satisfied with a solution without checking out the full ramifications it has for the system.

The first two are typical errors of novices, the last more of the expert. Experts are by definition the most capable of functioning at this level, but also the people who need to do so least often, because they have learned to reduce most problems to rules. If they have become, at the same time, less willing to accept that there are situations which do not fit their rules, they will be not be capable of shifting flexibly between the different behavioural levels to match the demands of the changing situation.

Another aspect of this problem with boundaries between levels of functioning is when people consciously set out to find other ways of achieving their objectives outside the routines which the system designer had in mind. Operators are endlessly creative in finding short cuts, many of which are perfectly safe and which the system designer would be only too glad to adopt, but some of which are insufficiently thought through because of insufficient system knowledge or time. These violations of system rules, as seen from the viewpoint of the designer or manager, are seen as the essential freedom of action by the operator (who in them effectively plays the role of designer).

LESSONS FOR THE PRODUCT-LIFE APPROACH

This brief survey of the way in which people can predict or precipitate deviations, detect them and correct them can provide a number of lessons to be borne in mind in a total approach to system safety.

1. Lessons for designers

The first lesson is one which ergonomists have been hammering at for a generation, namely the link between the design and the operation phase of the life of the system. The operator or user has too often in the past been saddled with the impossible task of recovering the inadequacies of the designer. Machines which operate in ways which do not fit expectations, routines which are easily confusible, and tasks which have been left to the operator only because technology cannot yet take them over, are all examples of accidents planted like time-bombs in the system. Users are to be congratulated that they manage for so much of the time to operate safely despite them. The increase in feedback from users to designers and the heavier emphasis on product liability which have occurred in the last decade are welcome signs that this tolerance is reaching its limit. It might be good (if somewhat Utopian) to apply the rule that designer should be required to become operators or users of their products; that architects should live in the houses they design, process and instrument engineers operate in the control rooms they develop and managers wear the protective equipment they require their workers to use. This was a rule applied by early lift and bridge designers to be the first to travel in or across their creations, and was used successfully to concentrate the mind of the manager of some explosive works who were required to live on the premises.

A second lesson is related to the first. The designer who did have a spell as operator would soon realise that he or she wanted to modify or adjust the design just a bit and not sit down and operate within rigid rules which were not perfectly attuned to operating conditions. This might stimulate the notion that, as designer, he or she had the responsibility to take account of this very natural tendency of operators. Designs are not rigid things which ossify when they come off the drawing board. Therefore enough information must be provided to the operators to allow them to oversee the room for modification and not fall into unsuspected traps; in addition, predictable modifications which will lead to danger (like removing guards or defeating safety interlocks) should be made as difficult as possible. It is no defence to say that the ingenuity of the operator will always be greater than the predictive capacity of the designer. While this may be true, it is possible to do a great deal to make the effort of breaking the rules just too much to make it worthwhile.

A third lesson for the designer is to consider the user (and also the installer, maintenance fitter and demolition gang) as an active element in the system with all the built-in characteristics outlined above, and not as a passive follower of rules like the hardware. If the task given to the operators is not satisfying or in keeping with their capabilities they will modify it. A popular reaction to this among engineering designers is to try to eliminate the human totally from at least certain phases of the system by automation. If this is indeed possible it can be an effective way of increasing safety. The difficulty comes where it cannot be done fully and the operator is kept in the system "just in case" something goes wrong, or to do some simple task which cannot yet be automated. Designers should realise that they are dealing with a situation here where such compromises give the worst of both worlds. The operators will probably not be capable of intervening when the need arises because they do not have enough hands-on experience and their mental models are not sharp enough to swing immediately into problem-solving mode; in addition, if they have little to do the rest of the time, they will occupy themselves with something else, which may even be working out how best to modify (or sabotage) the system. Designers need to have the courage of their convictions and decide to eliminate the operator entirely (e.g., automatic subway trains) or should start out from the design principle of giving the operator a satisfying job and automating the rest or supporting the actions of the operator in ways which are helpful and not threatening.

In order to profit from all these lessons the designer needs to be very flexible and open. Attitude cannot solve everything, however; the designer must be equipped with the tools to carry out predictions of user behaviour and possible hazards. These are only slowly being systematically developed, and designers still have to rely too much on thinking themselves into the position of users. This is difficult because designers are by definition experts and are also not inclined to think of misuse of their pride and joy. There is a case for deliberately involving outsiders in that stage to ask the questions that the designer is too close to the system to ask (e.g., "what happens if I press this and then turn that?"; "what would happen if I used your pruning shears to cut up my chicken joint?", etc.). If the work of psychologists such as Hudson (1967) is a good guide, there is also much to be said for not entrusting such creative, "what if ...?" questioning to people who have chosen the exact sciences as a subject of study because they cannot cope with ambiguity. Such people are poor at such predictions.

2. Lessons for system managers

Those responsible for running systems can also learn lessons from the description of the error making and correcting potential of humans.

The first lesson is to be suspicious of anyone who claims that safety is merely a matter of laying down and enforcing rules. This approach will go a long way, particularly the first element of it. It can never do any harm to define clearly and as exhaustively as possible how the system should operate to overcome all known hazards. But the second element is difficult to achieve. It will only work where it can be guaranteed that application of the rules will always result in safety, and it will work only with difficulty if following the rules is also not the easiest and most obvious way of doing the job.

Rules are subject to exceptions and to erosion. Safety manuals and safety laws tend to be full of complex specifications with many "if ..., then ..." clauses which are perfect if followed, but which are too complex to remember and for which execution of all the checks to see which sub-clause applies in any one case would take too long. It takes no skill to predict that they will not be followed and will only serve to assuage the consciences of those who made them by allowing them after an accident to establish exactly who should have done what and so who is to blame. The existence of such an edifice of rules is a signal that the system is inwardly sick and in urgent need of redesign to incorporate behavioural rules into either training or hardware design. Ideally design should precipitate the right action, and articulated, written rules are only necessary where the way someone would expect to have to operate in a given situation is not in fact correct.

Everyone has an interest in reducing the control of a system for the majority of the time to a rule-based level. It reduces the effort to carry out the task. What must be aimed for is that the user has made the rules (or fully internalised those made by the designer) and knows fully enough why they have been made to be able to modify them to cope with exceptions and with the dynamic environment of the system which will render old rules obsolete.

This conflict between establishing rules and leaving the flexibility to cope with exceptions and with changes can be seen at all levels in safety. It is reflected in the arguments about rigid central specification in laws and standards in contrast with enabling frameworks with objectives and the freedom for each company to comply in the way it wishes (Hale et al., 1990). It can be seen at the level of the company where operating managers are keen to reduce problems as fast as possible to routine rules in order to be able to get on with production. Safety departments have a task here to act as the protagonists of continual revolution in the firm (Hale, 1987). As soon as the rules have been laid down in a policy document and manual they should initiate the processes of monitoring and review which will eventually tear down that edifice and replace it with a new set of rules.

The final lesson for system managers is also related to the process and results of change. I have strongly advocated design and redesign as processes for solving safety problems. But they have their limits if they are confined to hardware design. The history of safety is full of the experience of redesigning a machine or system to remove one hazard only to find either that another has been introduced, or that the humans in the system use the redesigned system not to get more safety but to realise other goals (speed, saving of effort, quality, etc.). This phenomenon has been called risk compensation or risk homeostasis [for a review of the work of writers such as Wilde, Cownie, Calderwood, etc., see Hale and Glendon (1987, Chap. 9)] and is a simple restatement of what has been said above, that people learn to adapt themselves to systems and systems to themselves. It is foolish to expect that a system redesign will deliver the maximum expected revenue in extra safety if that expectation is based on the assumption that the people will go on behaving in the way they did in the old system. Straightening out roads to eliminate accidents results in higher travel speeds; putting extra automatic safety valves in a process results in more reliance by the operators on the system and a slower response to instances where the automatic valves fail; making chain saws lighter and safer opens up a new market for them among do-it-yourself enthusiasts who may be less skilled than professional foresters.

Part of this lesson is that people will quite happily trade some safety for other goals they care about. This is above all true of gains in safety where they were not initially particularly worried about the existing level. In such cases people can be expected to look actively for trade-offs, which must be actively counteracted if the safety gains are to be realised. As with automation to eliminate the human operator, design improvements in safety must have the courage of their convictions. If it is not possible to make the elimination complete, it may be better to leave an obvious and controllable danger which triggers the natural control reactions of operators, rather than making the danger invisible by hiding it behind screens (which can still fail) or by making it unlikely enough that it is forgotten about.

CONCLUSIONS

The object of this contribution was not to provide any easy answers to the incorporation of the human factor into a total systems view of risk control. There are none. The designers of many systems are working already close to the limits of controllability. A systems view does however highlight the room for manoeuvre which is created by trying to anticipate the activities of people in the various construction, use, maintenance and demolition phases of the product life.

The general message which I would like to leave is of the importance of feedback loops at all levels of system management in directing and learning from human behaviour. The individual uses them to decide which level of attention to devote to the task in hand and to learn from past experience what rules are appropriate and which may be eroded. The organisation needs them for similar learning processes related to redesign of processes which have gone wrong in the past and could not be fixed within the existing design. Active attention to enhancing and facilitating these feedback loops is one promising method of squeezing still more safety out of the systems which we have.

REFERENCES

Hale, A.R., 1987. Veiligheidsmanagement: Structuur, kaderstelling en aanpak (Safety management: structure, framework and approaches). Presented at the Symposium on Safety Management. Stichting Bedrijf en Overheid.

Hale, A.R. and Glendon, A.I., 1987. Individual Behaviour and the Control of Danger, Elsevier, Amsterdam.

Hale, A.R. and Pérusse, M., 1977. Attitudes to safety: Facts and assumptions. In: J. Phillips (Ed.), Safety at Work, SSRC Conference Papers 1, Wolfson College, Oxford, SSRC, pp. 73–86.

Hale, A.R., de Loor, M., van Drimmelen, D. and Huppes, G., 1990. Safety standards, risk analysis and decision making on prevention measures: implications of some recent European legislation and standards. J. Occup. Accid., 13(3): 213–231.

Hudson, L., 1967. Contrary Imaginations: a psychological study of the English schoolboy. Pelican.

Rasmussen, J., 1980. What can be learned from human error reports. In: K. Duncan, M.M. Gruneberg and D.J. Wallis (Eds.), Changes in Working Life. Wiley, Chichester.

Reason, J.T., 1987. Generic error modelling system (GEMS). In: J. Rasmussen, J. Leplat and K. Duncan (Eds.), New Technology and Human Error. Wiley, New York, pp. 63–83.

Comparative product-life-cycle confrontation of risks*

William W. Lowrance
Life Sciences and Public Policy Program, The Rockefeller University, 1230 York Avenue, New York, NY 10021, U.S.A.

ABSTRACT

Lowrance, W.W., 1980. Comparative product-life-cycle confrontation of risks. *Journal of Occupational Accidents*, 13: 47–54.

The health and environmental risks and benefits of technological products and processes should be evaluated and managed in comprehensive fashion, taking into account the entire production life-cycle from the obtaining of raw materials, to production, through distribution and use, and on through final disposal of the product and all associated wastes. We have begun to do this, in choosing among alternative sources of electricity, in developing liquefied natural gas systems, and in minimizing the risks in major industrial complexes. This ambitious technico-cultural foresightedness deserves to be encouraged.

INTRODUCTION

The basic precept of this conference is one that I strongly agree with; that is: The health and environmental risks (and benefits) of technological products and processes should be evaluated and managed in comprehensive fashion, taking into account the entire production life-cycle from the obtaining of raw materials, to production, through distribution and use, and on through final disposal of the product and all associated wastes.

This precept embodies an ethic of civic responsibility, and it holds many implications for engineering, management, and regulatory practice.

RISK FUNDAMENTALS

A definition. "Risk" is a compound estimate of the probability and severity of harm. Likelihood (probability) alone isn't sufficiently informative. Simply

*Presented at the International Conference on Industrial Risk Management, Zürich, Switzerland, 16–17 January 1989.

to say that harms occur more often in one situation than in another, for instance, neglects how damaging the harms are. Nor, obviously, is severity (magnitude) alone sufficient.

A key to dealing with risks is recognizing three separable modes of analysis and action:

Risk assessment is the description of the likelihood and severity of threat.

Risk appraisal is the evaluation of the personal or societal burden from the risk, and of the costs required for protection.

Risk response is the prescription of "what to do about" the risk.

These modes are distinguishable but interrelated. *Risk assessment* finds scientific facts; *risk appraisal* weighs consequences of the facts in light of personal and social values; and *risk response* makes action-decisions based on the facts, values, and pragmatics.

Appraisal is the element that bridges between scientific factfinding and social response. It estimates the risk burden, then it evaluates the payback from current or contemplated protective investments.

Subjects considered to be "at risk" may include any or all of the following: production workers, transport and other ancillary workers, plant neighbors, product consumers, members of the general population, nonhuman biota, and the inanimate environment.

If risk assessments are to be as broadly embracing as the basic precept of this conference requires, they must consider not just "normal" or "routine" risks (such as those caused by familiar materials failure), but also more unpredictable or quirky risks (such as those caused by damnfool human error, or by terrorism, or by odd coincidences of human actions with aberrations of nature).

COMPARING AND CONFRONTING

Here, I will state two "sermon" points that will be reflected in everything that follows.

First, we must be explicit and courageous in trying to *confront* – not just fear or gripe about – risks. Essential in this is accepting "as a fact of life" that some risks are a virtually irreducible price of maintaining central aspects of modern civilization. For too long we have avoided comparing risk to risk explicitly. And for too long we have allowed many decisions to be made de facto, unexamined.

And second, it is essential that we *compare* risks, in order to develop context and gauge importances. In this we must view risk-response actions not as wasted costs, but as investments that generate humanitarian returns. These returns include avoidance of pain, illness, incapacitation, death, disruption, and medical costs; longer and fuller lifetime contributions to society; reduction of environmental damage; and the securing of technological benefits as hazardous impediments are reduced. As with all investments, in general we will favor those that promise the largest return-on-investment. Approaching risk decisions as personal and societal investments is a powerful critical stance.

Being as explicit as possible (even if imprecise) about risk prospects helps set up the necessary confrontations; comparative appraisal helps resolve them.

EVALUATIVE SETS

Countless sets can be defined within which to evaluate and compare risks. These are illustrative:

Causative mode
mutagenic radiation
deafening noise
botanical allergens

Societal function
electricity generation
apple-crop protection
highschool athletics

Geographic area
neighborhood around a chemical waste disposal site
metropolitan Basel
alpine valleys

Environmental medium
drinking water
indoor air
sushi

Population
anesthetists
microelectronics assemblers
male Hispanic teenagers

Product class
snowmobiles
food colorants
contraceptives

Specified illness, disability, or mortality
cirrhosis of the liver
hearing impairment
spontaneous abortion

Thus, to understand the relative risk in medical X-rays, one might take stock of all the various ionizing radiation exposures that citizens normally experience, compare X-ray risks to other medical procedure risks, and assess the health risk in forgoing X-ray examination. Similarly, to guide effective design of programs to detect, prevent, and remediate the impairment of hearing, one might analyze the various risks to hearing (from various causes, to various kinds of people, at various ages).

Categories of this kind can be established within which to examine the full-cycle risks of various technological products and processes. The categories overlap (as with mutagenic risks from earthquake rupture of a nuclear powerplant).

FULL-CYCLE RISK ASSESSMENT

Although there were some less ambitious attempts earlier to examine risks broadly, probably the first large-scale quantitative assessments were performed in the 1970s in the energy arena. Others have followed. Even though each area inevitably has its own specialized complexion, the following examples will illustrate the overall notion.

Example: Electricity generation. Very elaborate studies have been made comparing coal-fired against nuclear-fueled power cycles. Most of these have examined: (a) the risks in fuel extraction, refinement, and transport; (b) the risks in construction of the powerplants; (c) routine operational risks to workers and plant neighbors; (d) rare but potentially catastrophic risks to workers and the larger population and environment; (e) human and environmental risks from fugitive plant emissions; (f) the risks in disposal of spent fuel residues and other wastes; and (g) the risks in decommissioning and dismantling the plants.

Some analyses have covered not only coal and nuclear but also other sources of electricity (such as natural gas, or hydro, wave, wind, or solar power). Still other analyses have compared electrical with non-electrical power systems (such as gas for heating).

All such analyses are extremely complicated. Their conclusions are quite sensitive to the underlying assumptions. Not surprisingly, many have been quite controversial (National Academy of Sciences/National Research Council, 1980; Inhaber, 1982).

Example: Liquefied natural gas (LNG). When, in the mid-1970s, major proposals were being made to establish liquefied natural gas ("energy gas") networks in Europe and the U.S., several generic analyses examined the risks in the technologies to compress (and thereby liquefy), barge, transship, store, and pipe this beneficially condensed but flammable and explosive fuel.

These analyses helped compare LNG with nonliquefied gas and with other sources of energy. But also they helped identify the high-risk elements involved. And the debates over the analyses led to the generation of criteria that should be considered when particular systems are designed or sited.

Some of these analyses were performed for corporations; some were performed for governments. A few, such as those developed by teams at the International Institute for Applied Systems Analysis in Vienna, probably exerted international impact, at least on the way other analysts conceptualized the problems (Kunreuther et al., 1982; Withers, 1988).

RELATED BROAD ASSESSMENTS

Over recent years several kinds of ambitious analyses have been conducted that relate to full-product-cycle risk assessment, and some of these have provided input to full-cycle assessments. Much of the methodology that has been developed through these exercises has broad application.

Example: Major industrial complexes. Although a few manufacturers had performed at least informal analyses earlier, probably the most elaborate publicly-discussed analysis of the overall risks associated with a large industrial complex was that conducted at Canvey Island in the U.K. in the late 1970s. Within an area of 39 km^2 on that island in the Thames near London are oil refineries, petroleum storage tanks, ammonia and hydrogen fluoride plants, and a liquefied natural gas facility. A thorough government inquiry analyzed the various accident risks to the island's thousands of inhabitants and workers. Then it made recommendations on how those risks should be reduced, such as by rerouting a road and building a blast wall (U.K. Health and Safety Executive, 1978; Woodward, 1987).

The Canvey analysis, which was probabilistic and based in part on the historic accident record for these and similar facilities, provided a powerful planning tool for the industrial and governmental institutions. And it provided an informative centerpiece around which the involved publics could debate.

Similar analyses have been conducted for chemical complexes in France, the Netherlands, and elsewhere. Some corporations have scrutinized their own facilities in this way, searching for unusually high-risk situations. And some have carefully analyzed process options, transport and storage options, and "cradle-to-grave" transformations of the materials they handle.

APPRAISAL OF RISK COMPLEXES

As is attested by the convening of this conference, we now are entering an era in which there is a demand to broaden the scope of assessments. This is a continuation of the movements, begun in the 1970s, toward "technology assessment" and "environmental impact assessment".

Apart from the examples mentioned in the previous section, not many extremely ambitious assessments have been completed. Nor have many entirely new technological enterprises been appraised early in their history. But a number of less broad, though important, appraisals are being pursued. For example, the risks in production, use, and recycling or disposal of packaging materials are being brought under scrutiny. So are the risks in manufacture, use, and environmental fate of agricultural pest-control agents.

In assessing the risks in any large enterprise, various aspects tend to be analyzed in technical detail separately, perhaps even at different times. But eventually a broad portfolio can be assembled and missing components filled in. Once the basic facts are in hand, the following steps can be taken to generate implications for action:

1. Describe the existing or state-of-art situation;

2. Transform the situation into scenarios, as dictated by various imposed chances (such as operational errors, or natural disasters) and control options;
3. Within each of the scenarios, appraise the various risks, the costs of control options, and the resulting (expected-value) risk burdens;
4. Take account of uncertainties and analytic sensitivities throughout;
5. Intercompare the major scenarios; then
6. Draw implications and set agendas:
   - for further risk research and assessment;
   - for improvement of controls;
   - for emergency preparedness;
   - for regulatory and other risk-management action.

A problem that could be examined by such a protocol, for example, is how to optimize a system of municipal waste treatment (including waste reduction, sorting, recycling, incineration, landfilling, etc.). The analysis would start with existing programs and facilities, project scenarios of improvements and possible new systems, and take account of potential error, pollution, and other such complications. It would appraise the risks, benefits, and costs. Then it would draw implications for improving trash collection programs, selecting disposal technologies, siting facilities, providing for disposal of ash or other ultimate residues, and the like.

ASSESSMENT AND APPRAISAL CRITERIA

Once the necessary technical and economic assessments are in hand, they are appraised according to criteria such as these:
- Risk to average exposed individuals;
- Risk to maximally exposed individuals;
- Risk to especially vulnerable individuals;
- Aggregate risk to the population(s) at risk;
- Proportionate risk within the larger (national?) population;
- Distribution of risk within or among various populations;
- "Weighting" of benefits for various affected populations;
- Cost per life-extension or health-preservation;
- Relative cost-effectiveness of various options.

As has been noted, the outcomes of all analyses are very sensitive to the boundaries that are adopted (such as which issues are considered) and the analytic assumptions that are incorporated. This is especially true for broad analyses, such as full-product-cycle assessments.

Typical bounding questions include:
Derivative effects: Consider secondary consequences? Tertiary?
Ancillary consequences: Consider risks in securing raw materials, etc.?
Subjects at risk: Be concerned about workers? Consumers? Plant neighbors?

General populace? Nonhuman biota? The environmental commons (stratosphere, antarctica, etc.)?

Societal structures at risk: Worry about employment? Disruption of the social fabric? Archeological or patrimonial inheritance?

Time horizons: How far into the future to look, for what effects?

Misuse or abuse causes: Anticipate consumer misuse? Worker error? Sabotage?

Associated physical causes: Anticipate nearby fire, firefighting? Railway or docking accidents?

Associated natural catastrophic causes: Anticipate flood, earthquake, mudslide, etc.?

The establishing of boundaries and assumptions is an ethical and political matter. One of the principal values of public review of analyses is debating these defining issues.

USES OF COMPARATIVE APPRAISALS

Optimization of screening, monitoring, or preparedness. Comparisons can indicate which adverse effects (illnesses, pollutant releases, accidents, etc.) should be watched for because they are most likely, or most severe, or both. And they can indicate needs for remediation or emergency preparedness.

Structuring of technological choices. Comparisons not only provide an array of facts, but they make explicit and help structure the making of choices. One of the ways they do this is by providing a centerpiece for debate.

Guidance of research and design. Comparisons can indicate what research should be conducted, and what design improvements need to be sought. When performed in a decision-analytical way, they at least indicate the relative "payback" that may be expected from the research or design programs.

Planning of risk-response actions. Comparisons help determine what kinds of risk reduction, accident preparedness, or other response actions have the greatest protective potential. Thus they help guide priorities.

ACTING STRATEGICALLY

More than being just an assortment of methodologies, full-cycle assessment and management amount to thinking and acting *strategically.* The essence can be summarized by a few simple-seeming but profoundly important rules.

Rule 1. Construe risks broadly. Use "scoping" appraisals to set priorities; use more detailed appraisals to evaluate the societal payback from specific risk-reduction options. Take all affected domains into account. Search for derivative and delayed effects. Suspect risks even in desired actions (such as worker risks incurred in removing asbestos from old school buildings). Think about the future.

Rule 2. Do not merely displace risks, or blithely substitute one risk with another, unexamined one. In reducing pollution in one environmental medium (such as air), do not in effect just move it to another (land and water). Do not blandly replace an existing product of known risk with a new product of unknown risk.

Rule 3. Overall: confront and compare. Be explicit. Do not let things "just happen". Examine options against each other. Work toward comprehensive approaches, based on clearly articulated rationales, promoted by broad social endorsement.

REFERENCES

Inhaber, H., 1982. Energy Risk Assessment. Gordon and Breach, New York.

Kunreuther, H.C. and Ley, E.V. (Eds.), 1982. The Risk Analysis Controversy: An Institutional Perspective. Springer Verlag, Berlin and New York.

National Academy of Sciences/National Research Council (U.S.A.), Committee on Nuclear and Alternative Energy Systems, 1980. Energy in Transition, 1985-2010. W.H. Freeman, San Francisco.

U.K. Health and Safety Executive, 1978. Canvey: An Investigation of Potential Hazards from Operations in the Canvey Island/Thurrock Area. Her Majesty's Stationery Office, London, U.K.

Withers, J., 1988. Major Industrial Hazards: Their Appraisal and Control. Wiley, New York.

Woodward, J.L. (Ed.), 1987. International Symposium on Preventing Major Chemical Accidents. American Institute of Chemical Engineers, 345 East 47th Street, New York, NY 10017.

Liability and compensation – the insurance and risk management point of view*

Orio Giarini

Association Internationale pour l'Etude de l'Economie de l'Assurance, Geneva, Switzerland

ABSTRACT

Giarini, O., 1990. Liability and compensation - the insurance and risk management point of view. *Journal of Occupational Accidents*, 13: 55–62.

The demand for liability covers is related to fundamental changes towards a "service" economy and is likely to grow substantially in the future. This development will require the most efficient redistribution of roles between government, industry and insurance.

1. THE DEVELOPMENT OF THE LIABILITY MARKET IN THE CONTEXT OF A CHANGING ECONOMY

The fundamental reason of growth in the demand for liability coverages and claims is related to the fundamental changes in the structure of the economy which is shifting from an "industrial" pattern to a "service" pattern. The clue to the difference between an industrial and service economy is rather simple:

- In a predominantly industrial economy the value of products resides in their mere existence (as John Stewart Mill said, their value is embodied in the product itself).
- In the service economy, the value of products and services is more and more related to their performance[1]. When the performance is considered inadequate, there is therefore a tendency to consider that the buyer or consumer has not received a fair value for his money. From this concept also derives the growing interest in the product-life concept of goods where in fact the optimization of their utility or utilization has to be considered, i.e. the period

*Presented at the International Conference on Industrial Risk Management, Zürich, Switzerland, 16–17 January 1989.

[1]See in this context: Orio Giarini, *The cultural foundations of the brand image of insurance*, the Geneva papers No. 48, July 1988, pp. 229–244; Orio Giarini and Jean Rémy Roulet, *L'Europe Face à la Nouvelle Economie de Service*, Presses Universitaires de France, Paris, 1988; Orio Giarini (Ed.), *The Emerging Service Economy*, Pergamon Press, Oxford, 1987.

of time during which they are useful and performing, as against the costs of the total product-life, i.e. the costs of the research and production of products, of utilization and maintenance as well as of their disposal.

The very nature of this conference can therefore be considered as a living proof that we are now increasingly operating within the context of a service economy. This is in essence not only true because of the growth of traditional services, but even more so because industrial activities today have to take into account the performance and functions of their products and tools and integrate a growing number of services within the manufacturing and distribution process itself.

The modern service economy can also be described as the phenomenon of servicialization of industry, similarly to what happened more than two hundred years ago when the first industries appeared and led to the industrialization of traditional agriculture.

Therefore, when we are talking today about the problems of liability and compensation we are in fact tackling a very fundamental issue in the production and utilization of wealth in the modern economy. The liability problems are not a passing issue, but a fundamental one which is here to stay. It does not concern insurers and risk managers only but, directly and indirectly, touches upon all problems related to the production of utilities in the economy.

After a discussion of this very fundamental starting point, we can then enumerate and comment some key issues related to the management of liabilities.

2. SOME SPECIFIC PROBLEMS AT STAKE

2.1 The Jury system in the United States

Confirming the remarks by Professor Baram at the Oslo Assembly of the Geneva Association (June 1985), the studies of the Institute for Civil Justice (ICJ) suggest that one should not assimilate the functioning of the Jury system as a whole, in the United States, with the specific excessive decisions which have hit the insurance industry. The studies of ICJ show that, with the exception of the cases well-known to insurance, the Jury system has not shown a clear trend towards a real increase of payments in favour of plaintiffs. Data available on the comparisons with professional judgements do not show general fundamental differences. It seems therefore that it would be improper to attack or criticize the Jury system as such on the basis of the excesses from which the insurance industry has suffered. Quite the other way around, those exploiting the present Jury system would then benefit if the system as a whole would be attacked. One should concentrate in those areas and specific cases which are clearly fancied by juries, and concentrate on the mechanisms producing the well-known economically unsustainable results.

2.2 The contingency fee system in the United States – The asbestos case

The large claims movement concerning asbestos is the fact of twenty-five lawyers who are operating in an almost monopolistic system, having investigated a lot in studying the problems and collecting every possible data.

In the last fifteen years, the asbestos claims have produced costs over one billion dollars, of which two thirds have been supported by the insurers. Total costs to defendants and insurers averaged about 95,000 dollars for average defense expenses, 25,000 dollars for average plaintiffs expenses and 35,000 dollars taken home by plaintiffs (37% of the total). It is here that one can see that legal costs are by far the major part of the problem (63%). One could even say that defense lawyers are more a part of the problem than a part of the solution.

2.3 The Jury system and the bargaining process in the United States

In the field of medical mal-practice, only 6% of the claims really go up to the stage of the verdict (and at this stage, 76% of the defendants are declared not negligent). Fourty-three percent of claims is settled before starting real litigation (one third of them with a plaintiff's settlement). Fifty-one percent starts a litigation bargaining process and stops before going to a verdict (of these, two thirds are settled in favour of the plaintiffs). All this shows that the verdict is per se only a signal and not necessarily always positive for the plaintiff, of a complex bargaining system.

It would be a mistake to criticize all professional lawyers in the United States, as the present difficulties are linked to the activity of a very small, although "efficient" minority.

2.4 The arguments of the social function of liability claims in the United States

It is often stated that the jury decisions and in particular malpractice, pollution and liability decisions, are taken in view of the low protection by the social security in the United States. If so, liability claims are:

(a) the most costly social security system in the world (over 60% to "administer" and "deliver" the system because of the lawyers' fee), and probably;

(b) a very unjust system because it only remunerates those who make a claim. An appropriate economic analysis could make clear that the "social" justification is untenable.

The social function of liability claims is obviously somewhat different from Europe. Comparisons between the advanced countries in the world are very difficult to make and sometimes utterly misleading. For this reason, the Geneva Association has recently started a preliminary study to compare the reasons, ways and results of comparable liability claims and compensations in some European countries such as Germany, U.K. and Holland, and the United

States. Preliminary results are published in "The Geneva Papers on Risk and Insurance" (July 1990).

2.5 The liability crises up to 1985 as a retarded cycle

The last insurance cycle in this sector came to a close four to five years later than expected, due to a situation of high interest rates and the development of a cash flow underwriting mentality, where the professional vocation of insurance (i.e., evaluate and cover risks) was somewhat mismanaged. This experience has also been a proof that elements exterior to the insurance industry and of a clearly general economic origin, are determinant in the development of the insurance sector.

2.6 Risk cover capacity from inside and outside insurance

In the United States pollution insurance is now fully excluded (with a couple of limited exceptions) and professional liability is subscribed essentially separately; excess policies are still rather restrictive. The difficulty to get liability insurance, added to the ups and downs of rates, has created an environment in which more and more risks are covered outside the traditional insurance market. Some experts have estimated that by the next decade, captives and pools will command 40% of the commercial market, representing approximately 25 billion of premium. The new risk retention groups in the United States and other initiatives like the bringing together of several captive companies, capable of facing some specific capacity problems are reinforcing the trend for insurance capacity being created outside traditional insurance companies. It would be interesting to examine if this trend will really change the structure of insurance supply in the long term: we think that at the end of this period, when the market will be more extended and established, the interest of all parties involved will be to maximize the efficiency of the risk distribution by a well integrated and collaborative insurance market. We can recall the historical experience when, for instance in the United Kingdom, some of the major insurance companies existing today, have been at the beginning the equivalent of modern captives, being created in periods of great changes in technologies and in the economic system.

The problem of capacity also has to be related to the question of how far risks are at all insurable. Beyond the upper possible level of insurability, the old question arises of the function of the state and of public institutions.

2.7 Controlling time – the claims made policies

Such policies have been presented for a time as being the key reference and strategy of the insurance companies in the United States and they are well justified because of the absolute necessity to control the period of liability. It would be, however, very interesting to find out how, in practice, these claims made policies work: problems have been mentioned as to the behaviour of the insurers and of the insured persons concerning the declaration of a claim with regard to their respective fears or interests of the reaction of their partners according to various moments at which the claims are declared. According to Dr. Pfenningstorf, who made a study for the American Bar Foundation (see 'Etudes et Dossiers' of the Geneva Association, Nr. 97, January 1986) "the introduction of claims made policy may become the final proof that the popular demand for broad and uncomplicated compensation can no longer be satisfied through liability insurance or through liability claims, but requires an all risks type first part compensation system under which persons suffering from injuries and diseases receive compensation without regard to the cause or causes. There is an important role to play for private insurers in such a system". This first party compensation system is seen as complementary to a residual tort liability market. Also the Institute for Civil Justice seems to consider the claims made policies as a solution of partial and limited effect for the future. It would probably be important to study in detail the future perspectives and alternatives in this area, keeping in mind the clear necessity to control the uncertainty of time.

2.8 Pain and suffering and punitive damages

This is again a typical American situation although some forms of compensation in this area are not always totally excluded in some European countries. According to ICJ, one should be able also in this area to distinguish the exceptional cases from a more general pattern.

2.9 The development of pools in the United States

They are taking place especially in areas such as pollution liability. But the main limitation is that pools always refer to a specific origin of the pollution. Criticisms are formulated because people suffering damages are paid only in those cases where the origin of the pollution is clearly indicated and has been foreseen in a pool. However, most cases of pollution do not have a very clear origin and even scientific knowledge cannot determine the origin in the majority of cases.

2.10 The technological scientific analyses of the origin of the damages

It has been suggested that "an Agency is needed acting for markets as a whole, having access to all scientific and technological sources, but acting solely in the interest of the insurers which could commission investigation into new risks to give underwriters the benefit of expert risk assessment". Increasing the ability for risk assessment has always been mentioned by insured as well as by insurers, and has been promoted by the Geneva Association for years through its activities in the field of risk management studies. But this problem is not always correctly discussed at two key levels:

(a) there is obviously a gap to be covered between what insurers should know concerning new technologies and products in various fields. This problem, with adequate research management, could be solved;

(b) on the other side, achieving the best possible knowledge does in most sectors not mean that the residual uncertainties which are implicit in the modern technological development can be eliminated. We go back here rather to "legal" risks, based on socio-economic pressures.

2.11 The new European directive on product liability

It shows that, although there is nothing comparable between the European and the American situation, the areas covered by different forms of liabilities respond to a general modern economic problem. If the situation will be properly managed in the future, it is clear for many experts that the liability market is an extraordinary important future market for insurance. Among the first criticisms to the EC directive, we can note: the limit of 70 million ECUs for claims, which is too high for small companies and probably too low for large international companies. The article of the directive concerning the "level of current scientific knowledge" is also considered by some as susceptible of diverging interpretations in the future as was the case in California.

2.12 IBNR (reserves for damages incurred but not recorded)

A key issue is finding the proper level of IBNR and the constraints for insurance and, in particular, of reinsurance. Whatever the type of insurance institution it is obvious that any risk management in the world inside or outside insurance has to consider this aspect of reserving as a key issue and manage it properly.

2.13 Liability and the economy

It seems rather obvious that economic studies in this area are largely underdeveloped. It would seem important to study in detail and with adequate facts the following question:

- Who really pays and how much for a liability claim? This question requires to examine how the consumers (buying the products of a liable producer) and the insured are hit by the raising cost-waves in the liability markets.

- It would be essential to make an economic analysis of the costs and efficiency of the legal system in providing justice in this area.
- In the case of the United States, one might even explore how far the liability situation is a non-tariff-barrier to trade and, as such, negotiable at the international level, for instance, at the GATT.

2.14 Integrating risk management

Finally, we would like to emphasize again the main subject of this conference, on which many other speakers will insist in more detail. It concerns the question of integrating the knowledge of the problems of risk management and its costs at several levels already at the beginning of any research and development programmes leading to new products, and therefore leading to new types of possible liabilities. In doing this, it is absolutely necessary to have a minimum level of adequate understanding of how the insurance industry operates and of the main problems faced in identifying, controlling and reserving for risks.

3. ADDITIONAL NOTES

3.1 Pollution compensation laws and the difficult balance between state incentives and legislation, industrial and insurance obligations

Recent negotiations in Sweden and Germany, show that, particularly in the field of pollution liability, a delicate equilibrium has to be found between the political objectives of governments as to protection against pollution and the role that industry and insurance have to play.

The Swedish government has issued a new legislation for pollution compensation which came into force on 1 July 1989. It comprises about 7000 companies in Sweden registered according to the pollution law. The claims are handled by a syndicate of five Swedish insurance companies.

This scheme followed a long discussion about the possibility of establishing a pure government fund, based on contributions from industries, or to introduce a compulsory insurance. This insurance cover has a limit of a total of 200 million Swedish Kronen per year and a limit of 50 million Swedish Kronen for each property damage. For personal injury, the limit is 100 million for injuries occurred over one year.

It is interesting to note that this scheme is based on a kind of claims made system, whereas all damages incurred and/or caused before the insurance contract are excluded.

Concerning Germany, the Ministries of Justice and Environment have proposed that "potential polluters with high risk will have to prove that they have liability insurance to get their operating licences".

This proposal has been criticized by both, industry and the insurance sector on several grounds:
- Most industries in Germany are already covered in such areas.
- There is larger disparity in the type of pollution various industries can produce and therefore in the type of liability they can provoke.
- Insurance is very reluctant to be a kind of government hand in the process of obtaining operating licences for industry, although in a sense this already happens for automobile compulsory insurance. But there is obviously a great deal of difference in the two cases.

In any case, what one needs to evaluate are essentially the following problems:
- Is the pollution liabilities issue of such a nature that it will become greater and greater or is this type of preoccupation bound to somewhat scale down? The answer to this question will obviously have an important effect on the political attitudes.
- In the case that the pollution liabilities will produce more and more problems and the necessity to find appropriate compensation systems, the question of the degree, quality and type of government intervention will become more and more crucial:
  - Will this be limited to a revision and eventually a hardening of rules concerning control and disposal?
  - Will this increase the role of governments as a kind of insurance institution using the fiscal tools (through a levy among all those who are polluters in principle)?
  - Will the governments try to stimulate the insurance industry more in the way they are doing in Sweden or Germany?
  - Will they simply create incentives (of which type?) so to stimulate insurance and industry to have as good covers as possible to compensate pollution liabilities of all sort?
- In addition, there is always a problem concerning the question of catastrophic damages[2], whereas the government is in any case involved above a certain level of losses. One should understand that governments will be interested in increasing the control or the self-control of industry and insurance at a lower layer hoping that this will also affect a better control of the upper layers of costs.
- A key issue never to be forgotten in any situation is the moral hazards effects of any form or scheme. A fair amount of personal responsibility must be preserved in all cases so to optimize the economic performance of any solution.

---

[2]See *The Limits of Insurability*, the Geneva papers on Risk and Insurance, No. 39, April 1986. For the same type of problems see also the particularly well written and documented article by William Thiela on Assessing the role of insurance in the commercialisation of space, in: *American Enterprise, The Law and Commercial Uses of Space, Vol. 3,* National Legal Center for Public Interest, New York, 1987, pp. 137-164.

The car – a challenge in functional requirements complexity, safety and reliability*

H.-H. Braess and G. Reichart

Bayerische Motoren Werke Aktiengesellschaft, Munchen, F.R.G.

1. INTRODUCTION

In the early days of motorization the driver was proud, to set his vehicle in motion, to compete his noisy and bumping vehicle with the horse-waggon on bad roads and to demonstrate his intimate expertise of the technique in case of disturbances (Fig. 1). Nowadays completely different criteria are valid.

The car as a subsystem of the traffic system and the car itself constituted by many sub-systems and components has to meet a huge number of requirements (Table 1). Many of these are fixed by laws and regulations. Therefore many conflicting goals arise. This development has not come to an end yet neither in qualitative nor in quantitative terms.

For a car a lot of potential designs and variants can be realized to meet the various requirements. A manifold of physical effects, function principles, constructive designs and systems are available. Even this spectrum is constantly increasing (Fig. 2).

The car consists today of 10,000 up to 20,000 parts and more than 100 materials, nearly the whole spectrum of manufacturing techniques is applied.

---

*Presented at the International Conference on Industrial Risk Management, Zürich, Switzerland, 16–17 January 1989.

Fig. 1. From the horse-drawn carriage to the automobile.

TABLE 1

Basic requirements made of an automobile (from the customer's perspective)

Technology	Economy	Pleasure and attractiveness
Transport capacity	Purchase price	Fitments
Active safety (avoidance of accidents)	Fuel consumption	Styling
Passive safety (avoidance of accident consequences)	Cost of maintenance	
Usefulness	Service	
Comfort		
Environmental care		
Quality, service life, maintenance requirements		

Nearly each design-parameter affects several attributed and on the other hand each attribute influences several design-parameters (Fig. 3).

Many different methods can be applied to the development of automobiles. The development process itself is characterized by a great number of work-packages, milestones and iterations. The task of automobile design and development has changed from a task, which could be solved in the early years by constructive ideas and empirical methods, to a task which nowadays can be called a "hyper-task" of the "mastering of an n-dimensional complexity".

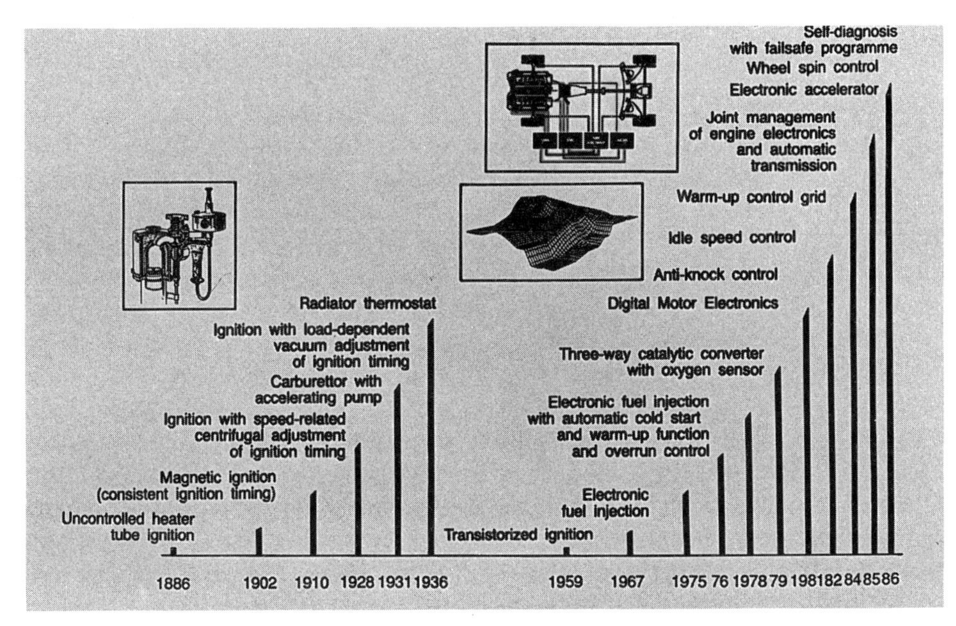

Fig. 2. Major milestones in the development of drive management and control systems.

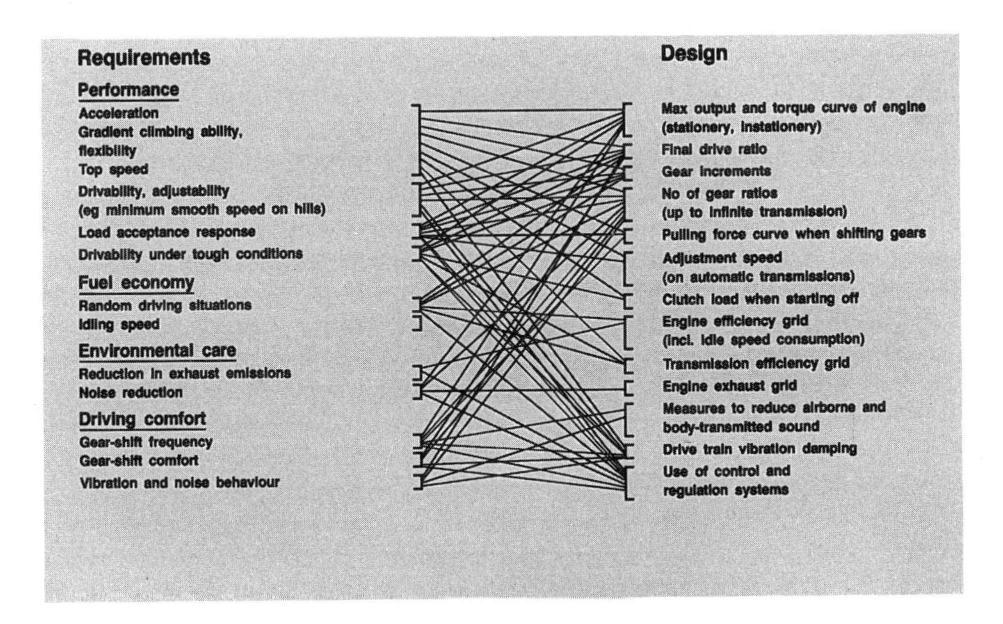

Fig. 3. Requirements and design of the engine/transmission system.

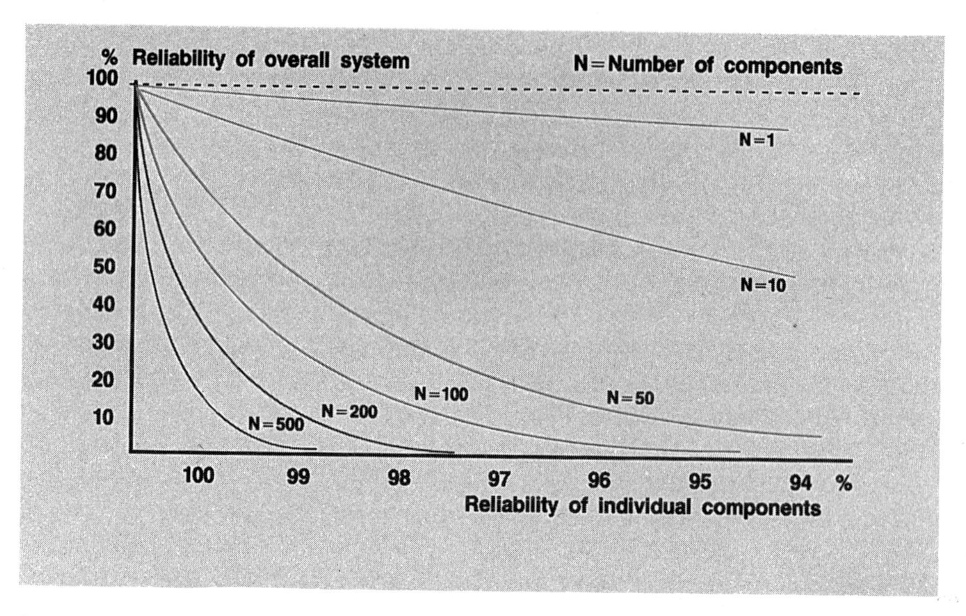

Fig. 4. System reliability as a function of the number of components and their individual reliability.

TABLE 2

Minimum random sample required for detecting more than 0 defects

Share of defects in total volume	No. of parts to be tested (minimum size of sample)
10%	23
1%	230
0.1%	2300
0.01%	23000
.	.
.	.
.	.

Note: Estimate based on a binomial distribution, 90% probability.

Particular emphasis has to be put on the reliability, availability and safety requirements, since even the best function is of little worth, if one cannot rely on it, or if it would lead to a dangerous situation. The reliability of a system, in which the components are functionally arranged in a serial manner, is decreasing with the number of components (Fig. 4). The system "automobile" constructed by a great number of individual parts requires therefore a very high quality and reliability of its components. This requires an enormous effort in development as well as in manufacturing, especially for systems with safety relevance (Table 2). Since the product "automobile" has to be attainable within

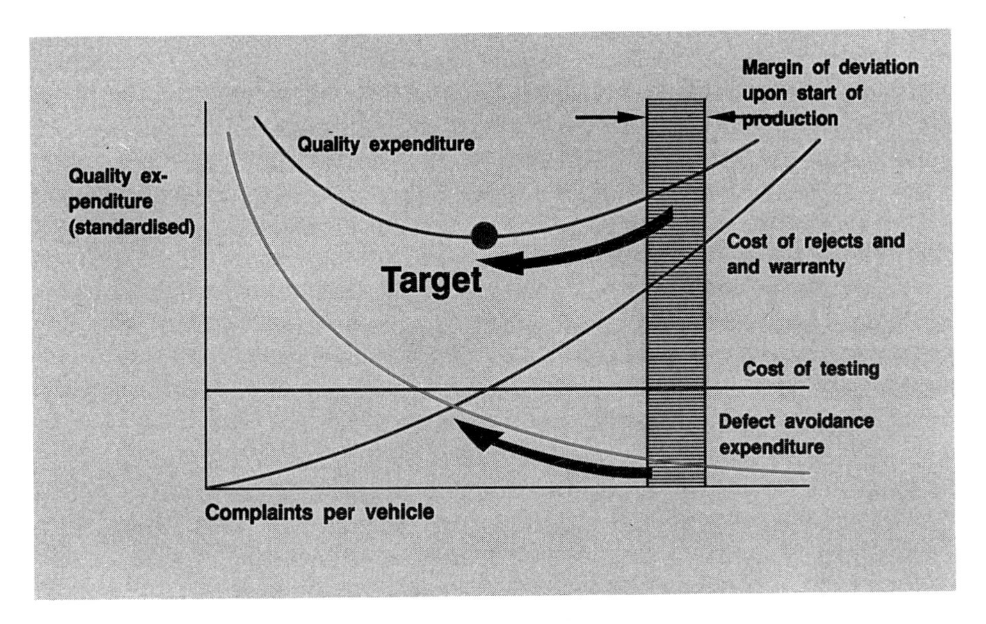

Fig. 5. Optimisation of quality expenditure.

the means of the potential customers, an optimization of the technical solutions with respect to costs is inevitable in the industrial practice (Fig. 5).

In this presentation some of the goal conflicts and critical issues for the design and development of automobiles will be discussed in more detail. Furthermore, an attempt is made to illustrate, how the increasing complexity can be coped with a systematic application of advanced, computer-based development and manufacturing methods and how the goal of a further improved safety and reliability can be achieved at acceptable costs.

2. EXAMPLES OF GOAL CONFLICTS AND CRITICAL ISSUES

A first example for the interaction of construction-parameters with various attributes of an automobile is the Driving Behaviour, an important factor of the active safety, that means the prevention of accidents. Optimal driving behaviour requires both the keeping of the desired course (Driving Stability), as well as the ability to follow quickly necessary steering actions (Steerability) (Table 3).

Important design parameters influence both requirements in an adverse manner. Therefore, a good compromise has to be sought carefully. A potential to alleviate the finding of a good compromise lies in the application of control systems as the classical example of *Antilocking-systems* shows, since control systems allow a flexible adaptation to the individual driving situation. However, such systems increase the necessary expense in development and manu-

TABLE 3

Some fundamental influences on the driving behaviour of motor vehicles

Measure		Driving stability	Steerability
Wheelbase	↑	↑	↓
Centre of gravity (share of weight on front axle)	↑	↑	↓
Side-slip stability ratio rear axle versus front axle	↑	↑	↓
Steering stiffness	↑	↓	↑
Share of front axle in brake forces	↑	↑	↓

facturing and the complexity, that means costs, weight and unfortunately even the failure probability.

A further example shows a goal conflict between active and passive safety. To achieve optimal viewing conditions for the driver the field of view should be as large as possible, a requirement, which can be fulfilled by big windscreens and thin roof-pillars. The maintaining of the integrity of the passenger compartment, especially in case of an overturn requires on the other hand a stiff roof construction with relatively big pillars, a quite obvious goal-conflict with the avoidance of blindspots.

A basic and important goal conflict in the manufacturing of cars is caused by the fact, that many attributes can only be achieved or even improved by additional systems and components. This leads inevitably to an excess-weight which is in contradiction to the requirements of saving of energy and raw-materials, environmental protection and economy.

The light-weight materials and constructions still have limitations with respect to operational strength and reliability. On the one hand extreme high loads up to loads coming from misuse can occur in the operation of automobiles. On the other hand a certain variance of the impact strength has to be taken into account even for classical, proved material. A sufficient distance of the occuring loads from those, which can be tolerated, is inevitable to minimize the "design dilemma" as shown in Fig. 6. The application of light-weight materials, which already have disadvantages with respect to costs, is further hindered by this effect because of the initial uncertainty in the knowledge of strength-variances in series-production and long-term operation.

This problem leads to the following fundamental statement, which is also valid for the manufacturing. "Quality and reliability of highly stressed parts of components are higher the more tolerant the selected materials are and the

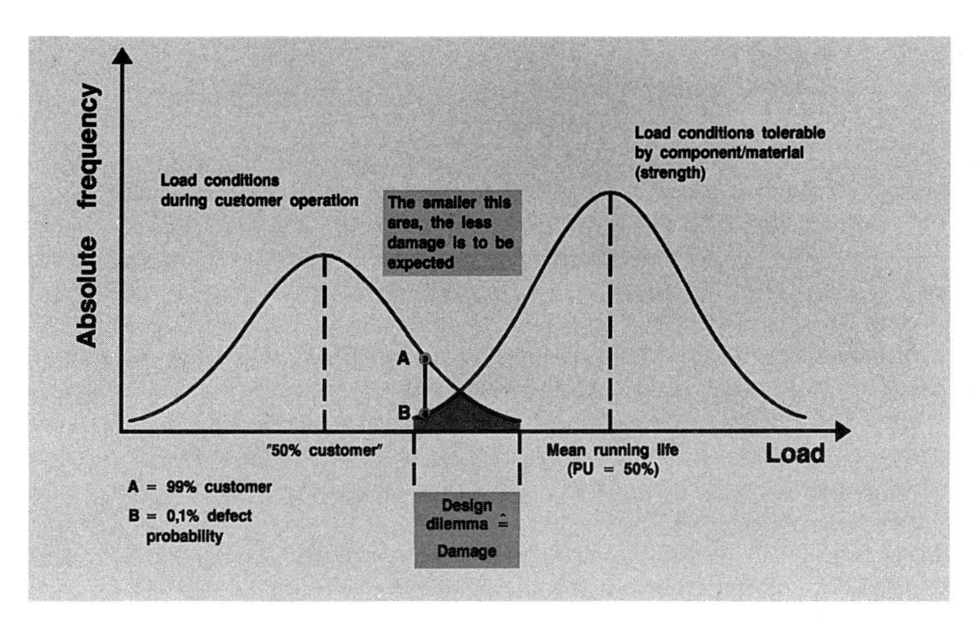

Fig. 6. Dimensioning risk as a result of inevitable deviations.

more the process-parameters and the stresses are away from the physical and technological limits" (Table 4).

The increasing complexity leads also to critical issues in the relation of the offered function to the operation and to test, maintenance and repair. The efficient use of unconventional systems has a prerequisite the acceptance of the users. This acceptance depends on various criteria, which are indicated in Table 5. This is especially true for new driver assistance systems, which will have impact on the drivers role. The driver assistance and information systems will either reduce the probability of occurrence or limit the consequence of human failures. Principal potential causes of human failures are shown in Fig. 7. Before the introduction of such systems many questions are to be answered. This is one of the goals of the European research project PROMETHEUS.

Complex systems require also corresponding measures for maintenance and repair. The customer's satisfaction with his car is greatly influenced by the amount of time and money he has to spend for maintenance and repair. Therefore, even from this point of view a high reliability and a low demand for maintenance are important requirements. The increase of the time interval between services during the last years has shown that a big progress is possible; further improvement can be expected. The BMW-service-interval display shows the necessity of a service depending on the actual strain. The service interval can be prolonged depending on the driving behaviour up to 5000 kilometers, more than are allowed by the usual fixed service interval without the risk of damaging the engine.

TABLE 4

Design and production: typical conflicts in interest

Typical characteristics of modern designs			Typical demands made of production technology
Application of sophisticated functional principles with many, in some cases highly accurate individual components	→	←	Not too many individual components
Maximum use of (invariably) highly sophisticated materials	→	←	"Easy" materials for simple and unproblematic production, simple joining methods
Relatively flexible parts with minimum tolerance	→	←	Tolerance not too small
Application of sophisticated, complex assembly processes	→	←	Easy-to-automate, i.e., simple and linear assembly processes, simple and straightforward assembly sequences
Simple adjustability of main design parameters	→	←	Avoidance of complicated adjustment operations

TABLE 5

Acceptance of unconventional systems by the automobile user: important factors

- Usefulness
- Transparency and consistency of system behaviour
- Tolerance against lack of practice and faulty operation, support in handling problems
- Compatibility with other systems, avoidance of risks
- Flexibility, individual adjustment to user
- Participation, freedom of choice
- Reliability

The more complex an automobile gets, the more difficult becomes the diagnosis of faults, especially in case of sporadic or transient faults. The increasing applications of electronic offer the potential to solve this problem by means of "on-board diagnosis systems". This technique allows the documentation and proof of occurred or only asserted malfunctions.

Expert-system-based diagnosis systems are apart from special applications still in a development stage. It can be expected that such systems will become

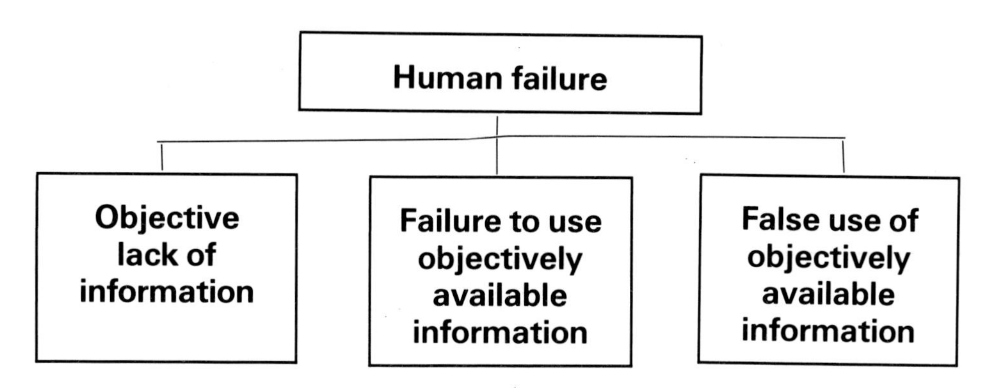

Fig. 7. Important causes of human failure (according to Hacker).

TABLE 6

Reasons for the application of expert system-based diagnostic systems

- Increasing system volume means increasing diagnostic requirement and greater data volume
- False diagnoses involve time and cost wastage as well as less safety
- Possible lack of fully trained human experts (not available at a certain time and/or place)
- Diagnostic process and its results reproducible and plausible ("self-explanation")

important for a better support of maintenance personnel and even in situations in which the expertise of specialists is not available (Table 6).

These examples have shown that the development and production of technical, economical optimal automobiles is a challenging task with respect to the applied technologies but also with respect to methods and tools for solving the goal conflicts and problems.

3. SOME ASPECTS OF THE DEVELOPMENT PROCESS OF ADVANCED AUTOMOBILES

3.1. Fundamentals

Figure 8 shows the usual development process of automobiles beginning from research over the early stages of development until the release to series production. The more ambitious the goals and the more innovative the concepts and technologies are, the more important gets a systematic development process. This begins with a proof of new materials, function or construction principles, and also includes the providing of solutions for all key-problem areas before the series-development starts.

Important tools for the support of the process of concept-finding and construction are:

(a) The application of comprehensive computation methods and tools (Computer Aided Engineering). General tasks and potentials of computation and simulation are shown in Table 7.

(b) The use of a great variety of experimental methods, which cover quite a large spectrum of different function and endurance tests (Table 8).

(c) The application of specific methods for quality assurance, reliability and safety management and for the planning of the maintenance and repair.

It is a characteristic of an automobile, that is is operated under a manifold of conditions with respect to driving, operation, road and environmental conditions (Fig. 9). Sometimes this environment could even be called hostile. Besides this some of the drivers misuse their car from time to time, either intentionally or unintentionally.

These influences have to be known in the development phase and hard but still realistic requirements for construction and tests have to be derived from them. The combination of various design methods and tools is very effective, because what the advantage of computation is at the one side of the disadvan-

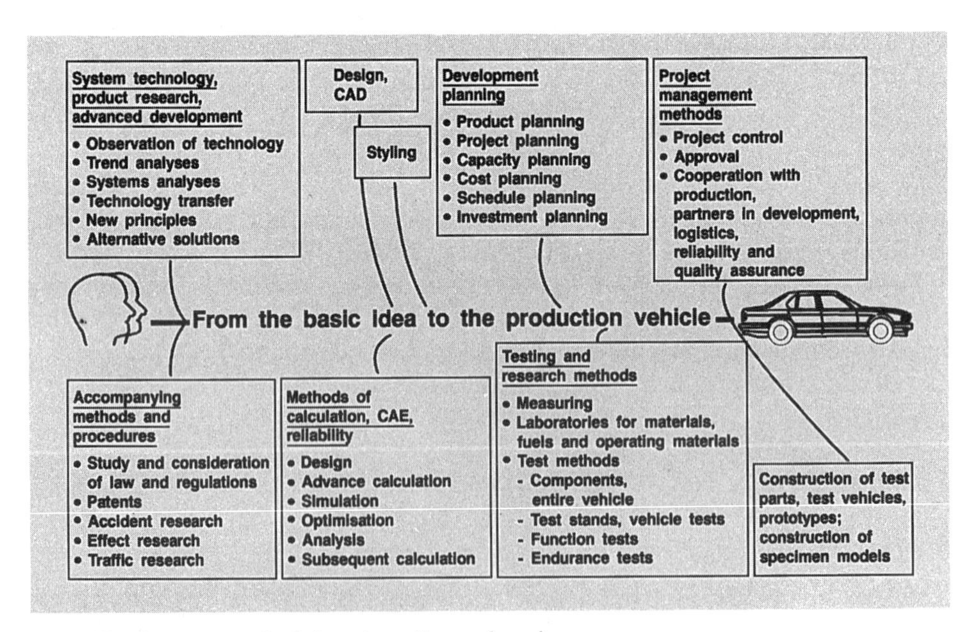

Fig. 8. Development methods in automotive engineering.

TABLE 7

Calculation: general tasks and specific issues

General tasks:

- Concept, advance calculation
- Optimisation
  - Structure
  - Parameters
- Analysis, subsequent calculation

Examples of specific issues:

- Dimensioning of aggregates and components (according to various criteria)
- Achievement of targets
- Coordination of systems, adaptation of sub-systems
- Minimisation of undesired effects (e.g. vibrations)
- Insight into the behaviour of complex systems
- Parameter sensitivities
- Worst-case studies
- Help in the interpretation of measuring results

TABLE 8

Testing methods in automobile development

Specific problems presented by the automobile:

Many different requirements and operating conditions and, accordingly, large numbers of tests, facilities and assessment criteria required.

Test methods required:

Function tests	- Long-term tests
Component tests	- Aggregate/system tests
Laboratory and test stand experiments	- Driving tests
Tests under normal conditions	- Abuse tests, accident simulation
Tests involving individual effects	- Tests involving combined effects
Tests with objective assessment criteria	- Tests including subjective assessments

---

tage of the test and vice versa. Even within one area of methods complementary and overlapping methods are used, as is indicated by the labor and field experiments in the field of operational strength (Fig. 10).

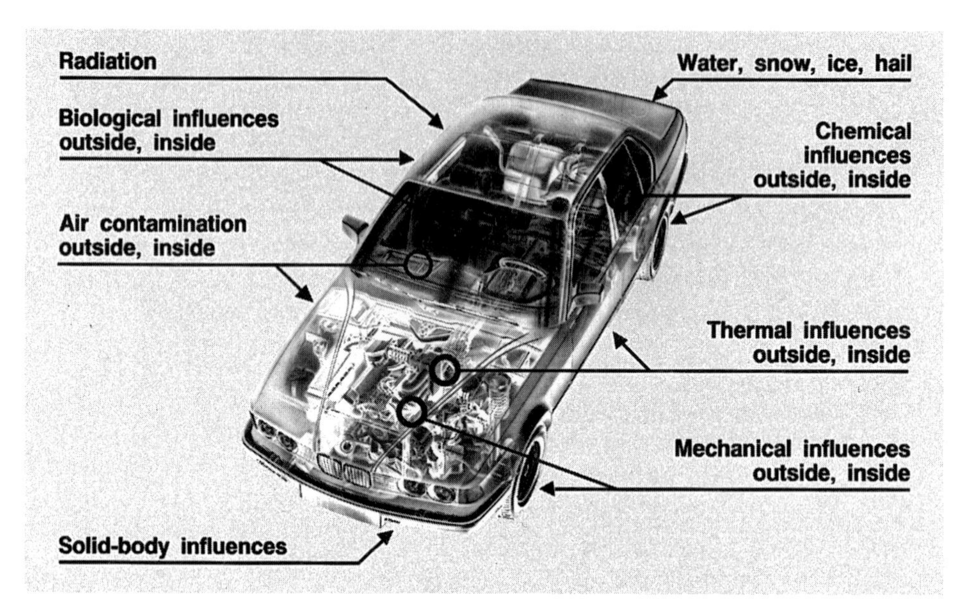

Fig. 9. Factors (specific load conditions) influencing the automobile.

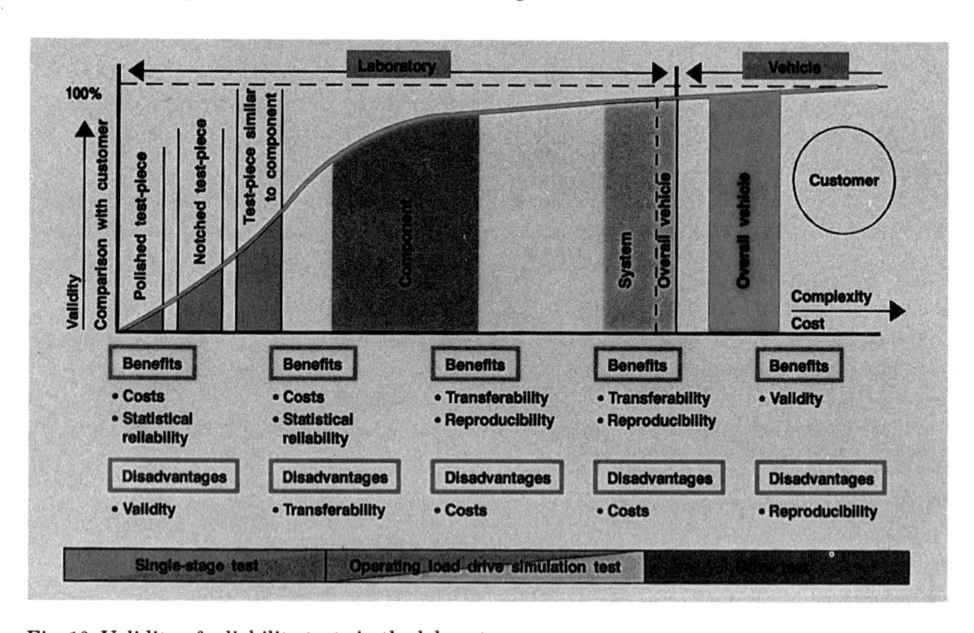

Fig. 10. Validity of reliability tests in the laboratory.

3.2. Methods to ensure quality, reliability and safety

Quality as the aptitude to meet fixed or presumed requirements, is nowadays nearly a self-evident characteristic of automobiles at least of the more expensive cars. This is also true with respect to reliability which is interpreted as quality over a period of time. Many activities for nearly all phases of product life have to be performed, as indicated by the quality-cycle according to Masing (Fig. 11).

Quality management, planning, control and testing are quite common amongst all car manufacturers. The reduction of production depth and the "Just-in-time supply" of components during the manufacturing require a close cooperation between supplier industry and the automotive industry with respect to quality assurance. Due to the increasing complexity of the supplied components the demand for capacity and qualification in the fields of quality testing and fault analysis is increasing. The quality assurance has to be accompanied by a high effort for achieving reliability and availability, especially of systems with safety relevance. Functions like ABS, Air bag, electronic gears, active suspension or electronic motor-management can have quite a big impact on safety, if appropriate measures are not taken (Table 9). This shows that reliability and availability are necessary, but not sufficient criteria. Protection against critical failures has to be achieved, e.g. by failsafe behaviour or the use of protection systems. Redundancy and diversity as important means to increase reliability have also limitations with respect to their application to mechanical systems in an automobile. This is due to functional reasons, space-, weight- and cost-problems.

This situation is different for sensors and for electronic systems, for which these means can be applied more easily than for mechanical components like

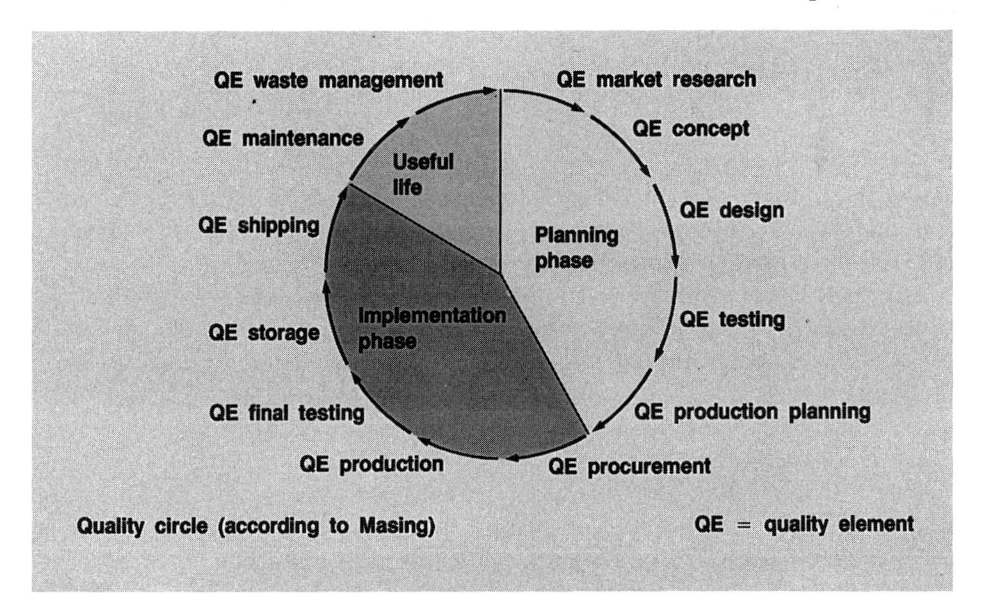

Fig. 11. Quality assurance throughout the entire product life-cycle.

TABLE 9

On the problem of active systems relevant to safety

System	Possible consequences of malfunctions
Anti-lock braking system	Failure of brake function
Airbag	Inadvertent actuation
Electronic transmission control	Down-shift from high speed Loss of acceleration powder
Active suspension	Unexpected change in driving characteristics
Electronic engine power control	Inadvertent acceleration Loss of acceleration

TABLE 10

Some fundamental procedures for enhancing the reliability of technical systems

Automotive technology	Procedures for improving reliability							
	Redundancy	Diversity	Pre-aging	Rust-proofing	Under-loading	Over-dimensioning	Maintenance	Wear-reducing procedures
Mechanical systems	•	•	•	++	++	++	++	++
Hydraulic systems	+	•	•	++	•	•	++	+
Electrics	+	•	•	+	+	+	+	+
Electronics	++	++	+	+	+	+	•	•

Key: • = small potential worthwhile, + = large potential, and ++ = very large potential.

actuators. For the latter methods like the use of high quality parts, fail-safe-characteristic, use of protection systems or appropriate maintenance concepts are required (Table 10).

The increasing application of electronic in automobiles requires that the traditional way of the add-on of new functions with individual solutions for the reliability aspects has to be changed. New integrated designs are necessary, likely in a configuration of distributed fault tolerant systems with pertinent distributed, multi-tasking operational software. An example for a potential structure of such an architecture is shown in Fig. 12.

Also a balance between the hardware and software reliability has to be achieved. The latter is a field of current research efforts. The experience in other fields of application, like aerospace, process or transportation systems will be carefully considered. It seems that methods for software specification and engineering, like EPOS, or methods for data flow analysis like STAN can be taken as a first orientation for potential solutions.

Safety, interpreted as absence of danger, requires moreover the assurance of a safe use of the systems. The systems should not cause danger by themselves and they have to be designed according to ergonomic principles so that a high human reliability can be achieved.

That is one of the reasons, why ergonomics play a very distinct role in BMW automobiles. The development of suitable system solutions has to be supported by analytical methods, which allow to compare different design concepts with respect to reliability and safety. Besides the well-established method of Failure Mode and Effect Analysis (FMEA) methods like Parts Count (PC), Reliability Block Diagrams and Fault Tree Analysis gain more and more importance (Table 11).

With respect to the aspect of safe use, intensive research is going on, to make a broad spectrum of ergonomic knowledge and tools including driver models available for the development process.

4. SUMMARY AND PROSPECTS

Some decades ago the Model FORD-T has shown that an extremely reliable car can be built up by a simple and robust construction and by a strict limitation of the scope of functions. In our pretentious era, the question can be posed: What is our situation like at the opposite boundary of the technological spectrum, as to say?

Even if everything offered by modern technology was realized, nevertheless

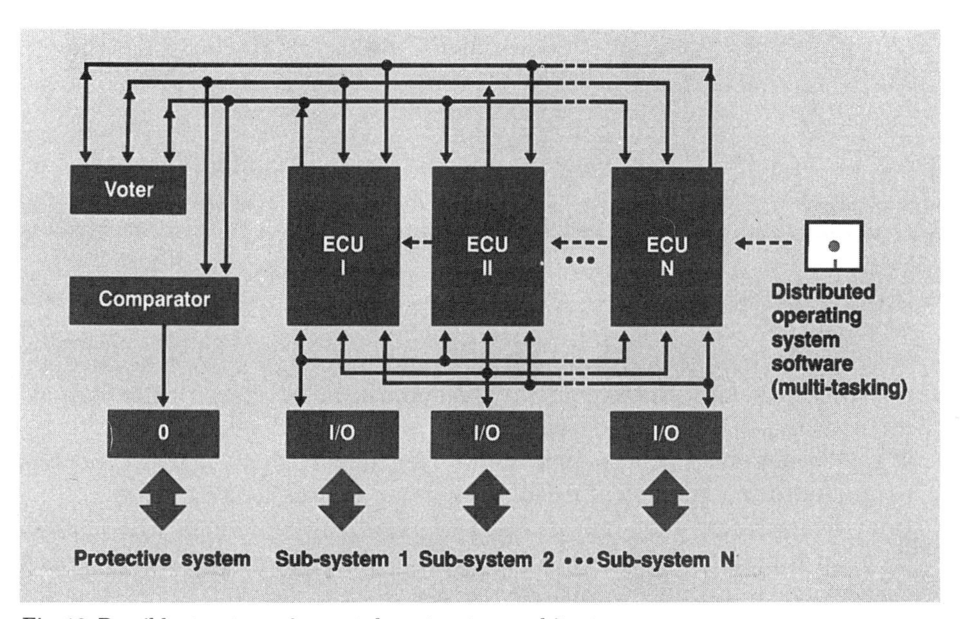

Fig. 12. Possible structure of error-tolerant system architecture.

TABLE 11

Conventional procedures in reliability technology

Method	Area of application	Objective	Input	Output
RDT	Definition and development phase	To examine system reliability based on the function interaction of components	Function scheme of system; if applicable, reliability parameters for function elements	System elements required for fulfilling function, degree of redundancy
PCM	Development phase	To determine the influence of parts on component reliability	Reliability data, number of parts used	Relative contribution of a part in terms of component reliability
FMEA/ FMECA	Development phase	To analyse the qualitative impact of different types of defects (components level)	Types of defects	Effects, criticality of defects
FTA	Development phase-production	To perform qualitative assessments on system level	Reliability data (Bool's representation of failure mode)	Failure probability, non-availability of systems
PARETO	Production phase-useful life	To assess cost/ value relationship of measure taken to reduce defect costs	Types of defects, defect frequency, costs	Defect cost: defect type ratio

one could not meet all requirements. Many of the goals conflicts are to severe. But even if there are no optimal solutions, nevertheless you can achieve good, even excellent solutions if a careful compromise between the various goals is chosen:

- in product design by a good combination of well-proved and innovative concepts and constructions
- in system technique by a comprehensive and systematic consideration of all relevant influencing factors and by the application of scientifically founded methods in planning and development
- in the development process and the whole product-life cycle by a synthesis of method-oriented and empirical procedures and last but not least
- from the technical–economical aspect by high-quality products with just the scope of functions, which is expected from the market and our potential customers.

Finally, the best chances in future will have a manufacturer who is able to master optimally the many goal conflicts. Quite often in technique the one has succeeded who has come from rather primitive solution over very complicated ones finally to an excellent simplicity.

Residual risk and its distribution in the project life cycle

A.S. Patwardhan, R.B. Kulkarni,

Woodward-Clyde Consultants, 500 12th Street, Oakland, CA 95418, U.S.A.

J.-F. Nicod

Woodward-Clyde Consultants, 1, Bd. de Grancy, 1006 Lausanne, Switzerland

1. INTRODUCTION

A typical product life cycle consists of three phases Design/Construction, Operation, and Decommissioning. Engineered structures are products with a long life cycle. Design/Construction periods of 10 to 15 years and operational periods of 50 to 100 years are common. Risk exists throughout the life of an engineered structure since the design and operation strategies seldom eliminate the risk completely. In the long life cycle, the magnitude of risk may change because the design and operating conditions may not remain constant.

Changes in the design conditions may occur due to changes in the loads and resistance of the structural members. Changes in the operating conditions may occur due to changes in the nature, frequency or intensity of use. Changes in the environment may change the character and size of those at risk. The combined effect may be a change in the nature and magnitude of risk during the lifetime of the structure. Moreover, the primary objective of the design strategy is to provide adequate structural resistance against loss of life in case of failure. The primary objective of the operational strategy is to provide a level of serviceability under normal conditions. Inspection followed by maintenance and rehabilitation are used to restore the loss of serviceability and/or structural resistance during the operational life of a structure. The changes in the design and operating conditions may occur either instantaneously as in the case of earthquakes or they may occur gradually as in the case of highway pavements.

There is uncertainty about the rate of change in the conditions and the timing of failure. Therefore, the management of risk during a long life cycle poses questions such as:

- What risks need to be managed?
- What approach should be used in assessing various risk mitigation strategies?
- How should the life cycle risk be distributed between the design and the operation/maintenance phases?

Further, the risk i.e., probabilities of losses of various magnitudes are influenced by choices made throughout the serviceability life of a facility or a product. These choices relate to the initial design of the facility/product as well as to monitoring, inspection, and maintenance policies. A facility may be designed with a higher margin of safety, which may reduce the frequency at which the facility is monitored/inspected and the frequency of maintenance/repair during the serviceability life of the facility. Conversely, a lower margin of safety may be incorporated in the facility's initial design with a plan to monitor/inspect the facility more frequently and thoroughly and maintain it as necessary. Thus, to maintain an adequate level of safety throughout the serviceability life of a facility, one may choose an appropriate combination of the margin of safety in the initial design and the frequency and scope of monitoring, inspection, and maintenance. This choice would be affected by such factors as the cost of increasing the margin of safety in the initial design, costs of monitoring, inspection and repair, characteristics of facility performance (for example, would the facility deteriorate gradually thus providing early indicators of loss of serviceability), and probabilities and consequences of failure. For a facility which is difficult to inspect and maintain and which may fail suddenly without exhibiting early signs of deterioration, an initial design with a high margin of safety would be more effective, since this would provide an adequate level of safety without having to rely on frequent inspection and maintenance. On the other hand, an initial design with a lower margin of safety may be optimal for a facility which deteriorates gradually and can be inspected and maintained with relatively small costs. An excellent example is the case of offshore oil platforms, which are expensive to build, difficult to inspect, are highly vulnerable to natural and man-made hazards and have resulted in service insurance losses.

This approach can be illustrated in the form of a decision tree (see Fig. 1). As shown in Fig. 1 different design strategies (such as strategies A, B, etc.) could be combined with various operational strategies of inspection and maintenance to maintain the structure at the desired level of safety or serviceability. Since there is uncertainty as to the operating conditions and the behavior of

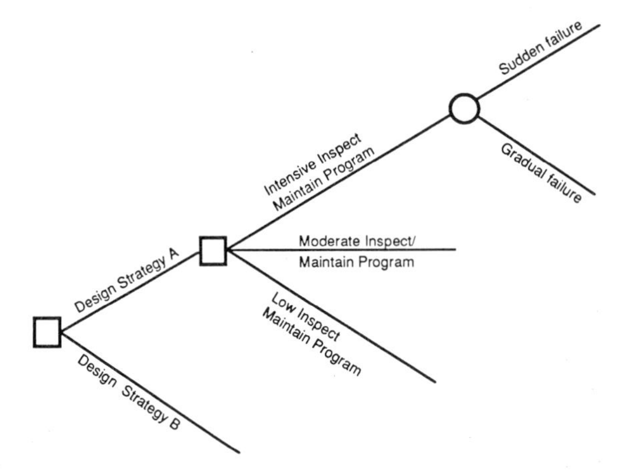

Fig. 1. Conceptual approach.

the structure through time it is not certain as to what combination of the design and inspection/maintenance strategies would provide the optimum choice.

Two different cases are examined:

a. A gradually disintegrating structure such as a highway pavement system.
b. A structure likely to disintegrate suddenly such as a building subjected to an earthquake.

Examples are presented to illustrate the approach in each case.

2. DEFINITIONS

Risks may be quantifiable, such as physical damage to a structure, or unquantifiable, such as hardship or loss of reliability of service. In the case of quantifiable risks, three different types may be defined (Patwardhan and Cluff, 1968). Calculated risk (R_c) is the estimated loss for a given probability of occurrence. Acceptable risk (R_a) is often based on a value judgment and may be defined as the loss(es) that can be tolerated without adverse consequences. Residual risk (R_r) is the difference between the calculated and acceptable risk and is the risk that needs to be managed. The relationship between the three risks may be expressed as:

$$R_r = R_c - R_a \quad (1)$$

Other definitions of residual risk exist in literature. For example, according to one definition residual risks is the difference between calculated risk and the risk mitigated either by insurance or through hazard reduction. In such cases, residual risk will by synonymous with acceptable risk. In a given area, factors contributing to the calculated risk (R_c) are the size, location, and frequency of causative events. Other factors being equal, the calculated risks may be lower for structures designed to modern codes, although the risks may not be eliminated entirely because the primary intent of codes is to minimize danger to life and not to minimize the disruption of functional use. Calculated risks may be constant or vary with time due to factors such as changes in structural resistance, property value, and increase in the population at risk.

Factors comprising the acceptable risk (R_a) are the amounts of loss that may be accepted through self-insurance, contingency planning, and other financial arrangements (R_{si}). The acceptable risk may be supplemented by external insurance coverage (R_k),

$$R_a = R_{si} + R_k \tag{2}$$

The magnitude of acceptable risk depends at least partly on value judgments, and may vary within wide limits depending upon the risk-taking posture of the decision maker. The acceptable risk may be small if there is no self insurance and no external coverage. On the other hand, it may equal or exceed the calculated risk if the combined coverage is high. In the case of certain important structures such as dams and nuclear power plants, acceptable risks may be governed by regulatory actions.

The residual risk (R_r) may vary widely depending upon the difference between the calculated and acceptable risks. It is low if the calculated risks are low and the acceptable risks are high. It may even be negative if the acceptable risks exceed the calculated risks, as in the case of vary conservatively designed structures; this may be the case for some nuclear reactors. Conversely, the residual risks may be very high if the calculated risks are high and the acceptable risks are small. An example is the case of certain older high-density urban areas in California where the calculated earthquake risks are high, the losses covered by self insurance are low, and the number of owners who have supplemented their coverage by external earthquake insurance is small (typically less than 20%). In such cases, the residual risk may be very nearly equal to the calculated risk.

The objective of a risk mitigation program is to reduce the residual risk (R_r) to a specific group or to the society. This can be accomplished by reducing the calculated risk (for example, engineering solutions to increase structural resistance, or land-use planning) or by supplementing the acceptable risk with additional risk coverage (for example, through insurance, financial incentives, or emergency procedures).

3. CASE 1: GRADUAL DISINTEGRATION

Pavements represent gradually deteriorating structures for which observations of advance signs of impending failure are possible. Pavements are designed to provide a structural integrity so that vehicles can travel on them safety and comfortably. Therefore, both structural resistance and serviceability are of importance. Once designed and constructed pavements undergo gradual degradation in resistance and serviceability. Most agencies collect pavement condition data on a regular basis to identify such signs. However, neither the timing of occurrences of these signs, nor the timing of actual failure following the signs can be predicted with certainty. Given this probabilistic behavior of pavements and the availability of periodic pavement condition data, a dynamic decision model is much more appropriate for such pavement management decisions as the selection of cost-effective pavement preservation actions and forecasting of future performance of a highway network. In the following sections the basic structure of a dynamic decision model are discussed, and a special class of dynamic decision models called "a Markovian decision process" are presented.

Decision models

A major objective of a Pavement Management System (PMS) is to assist highway managers in making consistent and most cost-effective decisions related to maintenance and rehabilitation of pavements. An integral part of a PMS is a decision model which can be used to determine the optimum type and timing of preservation actions for different pavement segments.

Two factors have a major influence on the choice of a decision model to be used in a PMS. First, the future performance of a pavement cannot be predicted with certainty. Thus, the behavior of pavements with time is probabilistic in nature. Since the future pavement condition is uncertain, the selection of a rehabilitation action appropriate for a given pavement at some future time is also uncertain. The second factor influencing the choice of a decision model is the periodic collection of pavement condition data. Most highway agencies conduct pavement condition surveys at some selected frequency (e.g., annually or biennially). The actual choice of a rehabilitation action at some future time can, therefore, be made based on the most recent condition survey. Since the planning period for any rehabilitation action is relatively short (generally less than 2 years), it is unnecessary and inefficient to choose a rehabilitation action for a given pavement several years in advance.

These two factors strongly suggest that the decision model for a PMS should be "dynamic", i.e., one in which the choice of a future action depends on the new information that would be available prior to making the choice. This is in contrast to a "static" decision model in which future actions are fixed at the present time based on present information.

Static and dynamic models

The major differences between the two types of models can be best illustrated by means of a simple example. Let us assume that the decisions of a rehabilitation action for a given pavement will be based on a single criterion namely, Present Serviceability Index (PSI). The minimum acceptable level of PSI is considered to be 2 for this example. Assume that the present PSI of the pavement is 3. Only three alternative actions will be considered: routine maintenance only, 1″ overlay, and 3″ overlay.

Static decision model

In a static decision model, future pavement performance following any of the rehabilitation actions is assumed to be known with certainty. Alternatively, only the expected performance is considered ignoring the possibilities of better or worse than expected performance. Hypothetical performance curves for the three rehabilitation actions are shown in Fig. 2. A major rehabilitation action will be selected for the pavement when it reaches the threshold PSI of 2.0. Thus, one rehabilitation strategy might be to apply 1″ overly at year 3, 3″ overlay at year 9, and 1″ overlay at year 19 (see Fig. 3). This strategy will maintain the pavement condition at or above the PSI of 2.0 during a selected analysis period of 20 years. Several alternative rehabilitation strategies can be defined. The total present worth cost of each strategy during the analysis period can be calculated including construction cost, maintenance cost, user cost, and salvage value. All the alternative strategies are then ranked based on the

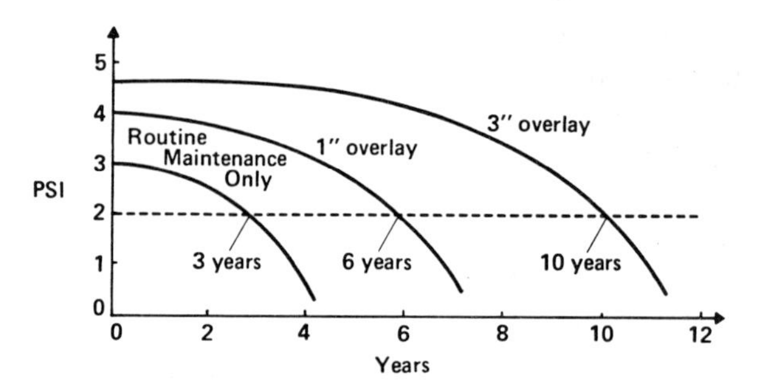

Fig. 2. Performance curve for the illustrative example.

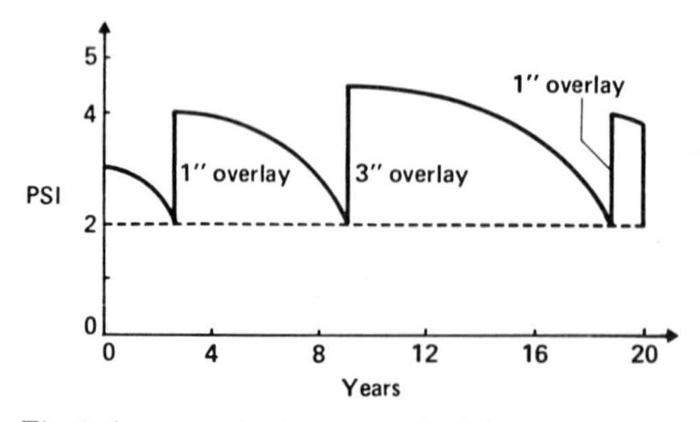

Fig. 3. An example of a static rehabilitation strategy.

total present worth cost and the one with the minimum cost is selected for implementation.

Some implications of this approach are worth noting. The choice of the action at the present time is strongly dependent on the actions selected for future time periods. Yet the future actions may not be taken at the designated time periods, because the pavement may perform better or worse than expected. This implies that not only the future choices of rehabilitation actions might be inappropriate, but also the choice of an action at present time could be ineffective.

For decisions under uncertainties, the expected cost is considered to be a rational criterion for ranking alternative courses of action (see, for example, Raiffa, 1968). However, a static model generally would not result in the least expected cost strategy. This is because of the nonlinear relationships between user cost and maintenance cost, and PSI. Thus, the cost calculated on the assumption of expected PSI behavior with time would not be equal to the expected cost of that strategy. In fact, it is likely that a strategy with significantly higher expected cost than some other strategy would be selected as being the best (the most cost-effective).

Dynamic decision model

Let us now consider how a dynamic model would analyze this problem. In this model, it is recognized that the future PSI following any of the actions is not known with certainty. However, probabilities of reaching different PSI levels as a function of time can be estimated.

Furthermore only the decision of what needs to be done right now is to be made at the present time. Decisions of future actions will be dependent (conditional) on the future performance of the pavement. The dynamic model can be illustrated in the form of a decision tree shown in Fig. 4.

A decision tree consists of two types of nodes – a decision node and a chance node, and several alternatives shown as branches at each of these nodes. At a decision node, the branches represent feasible alternative actions. The branches

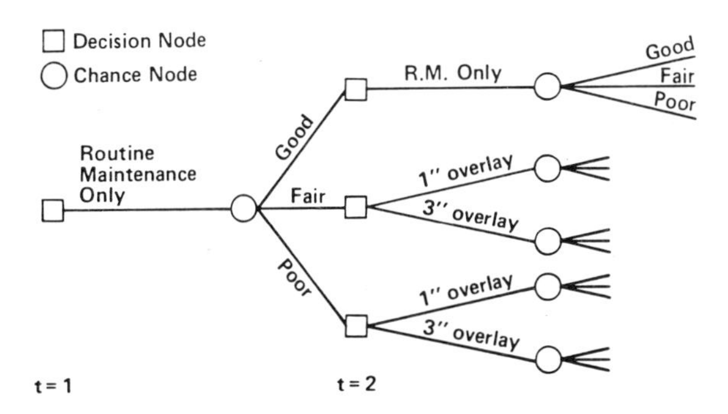

Fig. 4. An example of a decision tree.

at a chance node represent the possible outcomes of the action taken at the previous decision node. The probabilities of these possible outcomes are estimated.

Let us follow this structure for our illustrative example. Since the present PSI of the pavement is 3, the only feasible action is "routine maintenance only". This is shown as the only branch at the decision node at present time ($t=1$).

The pavement's PSI at the end of one time period cannot be determined with certainty. However, knowing the pavement characteristics, traffic and environmental conditions, probabilities that the pavement will be at different PSI levels can be estimated. For simplicity, let us consider three discrete levels of PSI – good (greater than 3), fair (2 to 3), and poor (less than 2). These three outcomes are shown as alternative branches at the first chance node in Fig. 4. Conditional on each outcome, appropriate alternative actions are selected at the beginning of the second year ($t=2$). For example, if the outcome is "poor PSI", the two alternative actions are 1″ overlay and 3″ overlay. Following each alternative action, the probabilities of three PSI levels are again estimated at the end of the second time period. This process is continued until the end of the analysis period is reached.

The analysis of a decision tree requires the estimation of probabilities and costs of different outcomes at each chance node. The costs would include construction cost, maintenance cost, and user cost associated with a given PSI level. The analysis is conducted by "folding" the tree backwards. Assuming n to be the analysis period, expected costs are calculated at each chance node at the end of the n-th time period. At the decision nodes at the beginning of the n-th period, the alternative actions with the minimum expected costs are selected. One then considers the chance nodes at the end of the $(n-1)$-th time period. Expected costs are again calculated assuming that the minimum expected cost actions would be selected at the following decision node.

The actions with minimum expected costs are again selected at the decision node at the beginning of the $(n-1)$-th time period. This process is continued until the first decision node is analyzed to select the action with the minimum total expected cost.

Note that the optimum strategy determined from a decision tree fixes the action only at the first time period. At each of the following time periods, the optimal actions are conditional on the possible outcomes at the preceding chance node. Thus, the optimum strategy might be identified as follows:

> Do only routine maintenance at $t=1$. If the pavement is found to be a good PSI level at $t=2$, then continue with routine maintenance only; if found at fair PSI level, select 1″ overlay; and if found at poor PSI level, select 3″ overlay.

The size of a decision tree can become extremely large for a real life problem.

This is because several distress types (instead of just PSI) may have to be considered separately in defining pavement condition and a large number of alternative actions may have to be evaluated at each time period. The problem is further complicated when a network of pavements needs to be analyzed to determine the minimum cost actions subject to the constraints of prescribed performance standards. In these situations, it would be impractical to analyze a decision tree by complete enumeration, i.e., by drawing all possible branches of the tree and evaluating each branch to determine the minimum cost actions. Fortunately, a special class of dynamic decision models, called "Markovian decision process" can incorporate several pavement condition variables and alternative actions, and also can analyze a large number of pavement segments. The details of this model are given in the next section.

A Markovian decision process

The problem of determining the optimum pavement preservation policies for a network of pavements can be formulated as a Markovian decision process that captures the dynamic and probabilistic aspects of pavement management. The main components of a Markovian decision process are condition states, alternative pavement preservation actions, and cost and performance of these actions. A condition state is defined as a combination of the specific levels of the variables relevant to evaluating pavement performance. For example, if pavement roughness and cracking were the only relevant variables, one condition state might be defined as the combination: (roughness = 50 in./mile and cracking = 5%). Note that the definition of a condition state retains the descriptions of individual pavement distresses and consequently, better matching of preservation actions to pavement condition is possible. This is in contrast with an alternative approach in which a combined score is calculated from the levels of individual pavement distresses. In the latter approach the specific causes of deteriorated pavement condition cannot be identified if only the combined scores are predicted for future time periods.

Alternative pavement preservation actions could vary from "Do nothing" to "Routine maintenance only" to minor and major rehabilitation. The performance of these actions is specified through transition probabilities. A transition probability, $p_{ij}(a_k)$ specifies the likelihood that a road segment will move from state i to state j in unit time (e.g., one year) if action a_k is applied to the pavement at the present time. A Markovian process is assumed to have only one-step memory. Thus, the transition probability is assumed to depend only on the present condition state i and not on how the pavement reached that condition state. Note, however, that by including factors such as age and design life of the last rehabilitation action in the definition of a condition state, the one-step memory can be made to consider the effect of type and time of the last action. A preservation policy for the entire network is the assignment of an action to each state at each time period.

Under the assumptions of a Markovian process, the specification of condition states and transition probabilities for alternative actions permits one to calculate the probabilities that a road segment would be in different condition states at any future time period for an assumed preservation policy (see Howard, 1971). The probability that a road segment is in a given condition state can also be interpreted as the expected proportion of all segments in that condition state. This allows the calculation of the expected proportion, q_i^l of the network of road segments in the i-th condition state at L-th time period for a given preservation policy. The performance of the network can be evaluated in terms of these proportions. For example, desirable and undesirable condition states can be defined and the proportions of the network in these two categories can be plotted as a function of time and whether the "health" of the network is improving or deteriorating can be assessed. A major objective of pavement management would be to find the preservation policy that would maintain desired performance standards over a long period of time at the lowest possible cost.

For a planning point of view, it is desirable that after some initial transition period, T, the network achieves a "steady state" condition. A steady state condition means that the proportion of road segments in each condition state remains constant over time. Mathematically, this implies that

$$q_i^T = q_i^{T+1} = q_i^{T+2} = \ldots, \quad \text{for all } i.$$

The advantages of reaching steady state is that the preservation policy will be stationary after time T, i.e., the selection of the preservation actions will be a function of condition state only and will not be affected by time. The expected budgetary requirements will also remain constant once steady state is obtained.

The user agency may desire to have control over the time, T, it would take for the network to reach steady state. Depending upon the initial conditions and the available budgets during the T time periods, short-term standards that are somewhat lower than the long-term standards may be acceptable. Optimal short-term policies (which may be different from the optimal long-term policy) can be determined to upgrade the network from its present condition to the long-term standards in time period T with minimum total expected cost while maintaining short-term standards during the first T time periods. Figure 5 shows the relationship between projected performance and budgets under optimum conditions. Mathematical formulations are presented in Kulkarni et al. (1982).

A successful application of the Markovian decision process was made for the highway system of the State of Arizona in the U.S.A. The heart of the Arizona's pavement management system is an optimization model, termed the Network Optimization System (NOS). It recommends pavement preservation policies

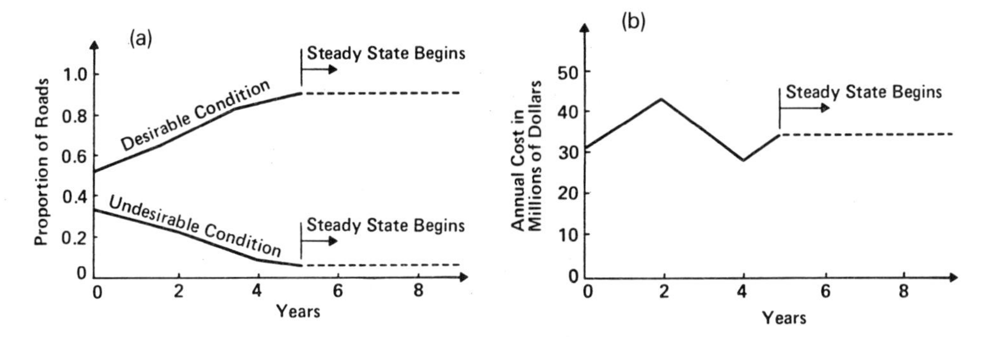

Fig. 5. Projected performance and budgets under optimum policies: (a) projected performance, (b) projected preservation.

that achieve long-term and short-term standards for road conditions at the lowest possible cost. The NOS is based on formulating the problem as constrained Markovian decision process that captures the dynamic and probabilistic aspects of the pavement management problem. Linear programming is used to find the optimal solution. Use of the model over a five year period between 1982 and 1987 is estimated to have produced a saving of approximately $101 million dollars. The details of this system are provided in Kulkarni et al. (1982) and Golabi et al. (1982).

4. CASE 2: SUDDEN DISINTEGRATION

An example of sudden disintegration is failure due to an earthquake. In this case it is not feasible to employ different inspection/maintenance strategies and the risk management methods used by public and private organizations include (see WCC, 1980):

(a) adoption of appropriate design standards to achieve a desired level of structural resistance.
(b) land-use planning to reduce the type and size of investment at risk.
(c) monitoring and emergency planning to reduce the extent of the post-event damage.
(d) financial incentives to provide for a favorable basis for faster replacement of existing construction.
(e) insurance to underwrite risks to existing and proposed construction.

These options are briefly discussed below.

(a) *Engineering design standards* consist of codes, design procedures, and regulations that are aimed at achieving a desired degree of resistance to minimize damage and life loss. This alternative has been the most commonly used option for reduction of hazards. For example, among the legislative actions in the United States related to seismic hazards, engineering solutions have been used by federal, state and local agencies in over 60% of the cases. To a lesser extent, emphasis has been placed on preparing standards for the construction and operation of facilities. The bases used for establishing the degrees of seismic resistance vary depending upon the importance of the structure, consequences of failure and the risk posture of the policy makers. Therefore, variable standards may exist for the same type of facility in different geographic areas. An example is the variation in building codes in different urban and rural local jurisdictions in California. Sometimes a formally stated policy does not exist and the basis for engineering design is a set of industry or professional guidelines. In other cases the acceptability of the proposed design standards is evaluated through a review of the possible consequences given in Environmental Impact Reports (EIR). If the estimated consequences are not acceptable to the reviewers, a change is required. Current policies for risk reduction on the basis of engineering standards do not require estimates of damage. Since most codes do not aim at total elimination of risk, a residual risk can be expected to exist in almost all cases. Therefore, this alternative for risk management may have to be supplemented with the use of other alternatives such as emergency preparedness or insurance.

(b) *Land-use planning* has been used as an instrument of hazard reduction policy by a number of agencies. The chief use of land-use planning in seismic risk reduction has been in controlling the type and size of development in areas susceptible to severe seismic effects, such as areas adjacent to active faults and slopes susceptible to failure. Control has been exercised in requiring special studies, special design features, and by limiting the number and size of structures. For example, recent legislation in California that addresses land use as a factor in policy formulation is the legislation to require counties and cities to prepare and adopt a seismic safety element as part of the local general plan, and the Alquist-Priolo Special Studies Zones Act of 1972.

(c) *Monitoring and emergency preparedness*. This element of risk management relies on monitoring of areas susceptible to earthquake effects and preparing plans for reducing the immediate post-earthquake losses. The measures are directed towards secondary effects such as fire, and consequential effects such as injury and life loss due to dam breach. Emergency preparedness is a useful supplement to other risk management elements to reduce residual risks.

(d) *Financial incentives*. Financial incentives are a form of risk management especially applied to accelerated retirement of existing structures. Incentives may include tax breaks, accelerated depreciation, or funding assistance in strengthening and reconstruction. Although it has been examined in some depth, the use of financial incentives is very limited. A related concept is one of financial disincentives in which penalties are imposed on the construction and operation of facilities in areas susceptible to severe earthquake effects. An example of a financial incentive is the program of federal grants under PL 93-

288 for the reconstruction and relocation of public buildings. Typical examples of where such incentives would be best utilized are existing hospitals.

(e) *Insurance*. Insurance is the transfer of risk to an underwriter for a reasonable cost. Again, as in the case of emergency preparedness, insurance is a supplemental form of risk management to manage the risks of both direct physical losses and consequential losses, such as loss of business and liability. Insurance is also used for trading off the cost of various design and operational strategies with the cost of underwriting the risk of damage. Its use has been guided by regulatory requirements and concerns for the protection of investment. However, the use of insurance as an instrument of public policy has not been explored to the fullest extent. Because continued operation through an earthquake is an essential performance requirement for most critical facilities, the chief purpose of insurance is to provide for consequential losses rather than for a major physical loss.

Because a program of seismic risk reduction must extend over a period of time, say a couple of decades, residual risks will continue to exist and policies will be needed to manage the risks. The policies should identify various risk management elements and their mix. Figure 6 presents a conceptual frame-

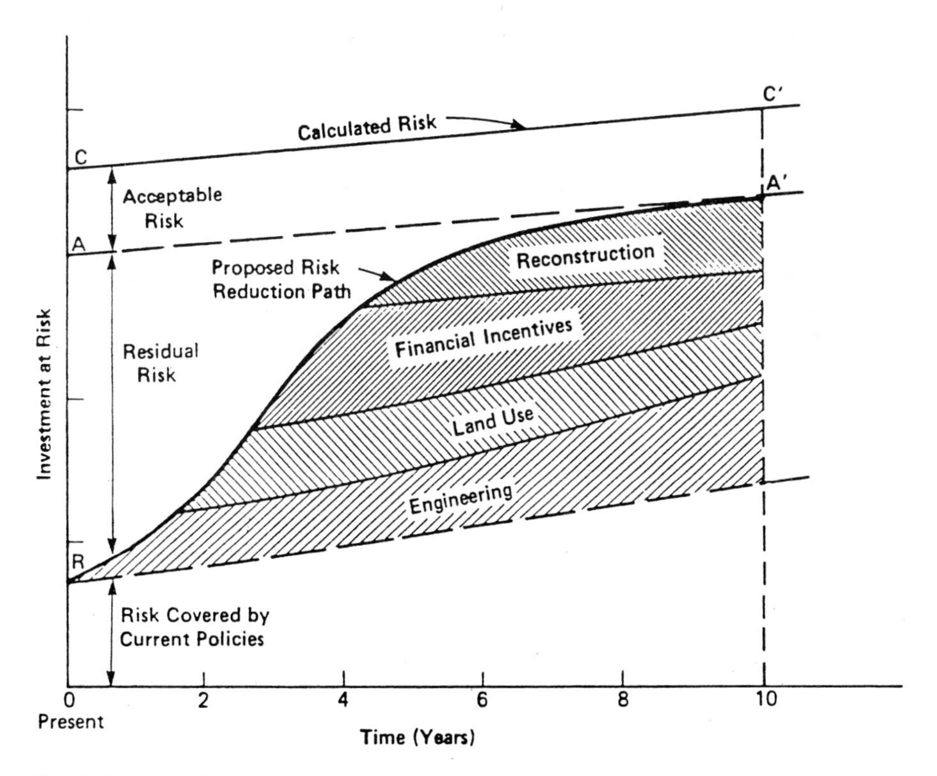

Fig. 6. Conceptual earthquake risk mitigation plan.

work for policy making for a typical facility. Assuming a 10-year period for achieving the policy goal, the figure shows a possible path for risk reduction and the component actions. At any point in time, the difference between calculated and the acceptable risks is the residual risk that needs to be managed. Details of the policy define the mix of risk management options: engineering solutions, land-use planning, emergency preparedness, financial incentives, and insurance. Based on the relative proportions of the various measures the cost of risk mitigation can be optimized.

The calculated and residual risks defined above provide a useful basis for risk management. Usually, risk assessments are desired for a single earthquake or for the combined effects of a number of events expected during a certain period of time. The following table shows the utilization of risk assessment for single and multiple events.

single event	(a) probability of different amounts of loss
	(b) probability of maximum loss on one structure
	(c) probability of different amounts of losses in a given area
multiple events	(a) risk exposure for an individual structure for a given time period
	(b) combined risk exposure for an area

Risk evaluations for single events can be utilized for emergency preparedness planning, for assessing insurance coverages required, and for assessing the total risk in an area. Risk evaluations for the combined effect of multiple events can be utilized for earthquake hazard reduction planning, for assessing frequency of insurance payments, and for emergency preparedness.

REFERENCES

Golabi, K., Kulkarni, R. and Way, G., 1982. A statewide pavement management system. Interfaces, 12: 5–21.

Howard, R.A., 1971. Dynamic Probabilistic Systems, Vol. I. Wiley, New York.

Kulkarni, R.B. et al., 1982. Development of a pavement management system for the Arizona Department of Transportation. In: Proc. 5th International Conference on the Structural Design of Asphalt Pavements. Vol. 1, pp. 575–585.

Patwardhan, A.S. and Cluff, L.S., 1978. The concept of residual risk in earthquake risk assessment. In: Proc. 2nd International Conference on Microzonation, San Francisco.

Raiffa, H., 1968. Decision Analysis: Introductory Lectures on Choices Under Uncertainty. Addison-Wesley.

Woodward-Clyde Consultants (WCC), 1980. Assessment of public policy regarding lifelines and critical facilities in California. Report for Seismic Safety Commission, State of California.

Concluding Talk

Consequences and visions*

Trevor A. Kletz
University of Technology, Loughborough, U.K.

INTRODUCTION

I summarise below some points that have already been covered and some that have had less attention than they deserve.

FLIXBOROUGH: THE GREEN RESPONSE

In my first talk I said that the explosion at Flixborough in the U.K. in 1974 had produced an explosion of papers on ways of preventing similar incidents happening again. Most writers had suggested adding on to our plants protective equipment to *control* the hazards. I suggested instead that we should look for ways of *removing* the hazards and I summarised some actions that could be taken. A third point of view, what we would now call the green response, shows how the public's perceptions of risk can be incorrect.

Flixborough was described, correctly, as the price of nylon. If the price is 28 people killed why do we not go back to the ways of an earlier generation and use wool or cotton instead? However, agriculture is a high risk industry and more people will be killed making a billion wool or cotton garments than the same number of nylon garments. Flixborough was the price paid for nylon (not an inevitable price) but the price of wool and cotton is higher.

TEACHING LOSS PREVENTION TO STUDENTS (Kletz, 1988a)

To be able to join the U.K. Institution of Chemical Engineers as a full (or corporate) member it is necessary to have a degree in chemical engineering from a University or College which follows an approved syllabus (though courses which, for good reasons, diverge in detail from the syllabus may be accredited) and at least three years experience. The syllabus was last revised in 1983 (a further revision is in hand) and now includes safety and loss prevention. This U.K. practice is unusual. In most countries, including the U.S.A.,

*Presented at the International Conference on Industrial Risk Management, Zürich, Switzerland, 16–17 January 1989.

the majority of chemical engineering students get little or no training in loss prevention. Why then do we in the U.K. think that loss prevention is an essential part of the training of chemical engineers? There are several reasons:

1. Loss prevention should not be something added on to a plant after design like a coat of paint (though sometimes it is) but an integral part of design. Hazards should, whenever possible, be removed by a change in design rather than by adding on protective equipment. A chemical engineer need not know much about paint – an expert on paint can tell him which sort to use – but he should not leave the safety of his plant to the safety officer. If he does so the safety officer will add on protective equipment of various sorts – trips, alarms, fire protection, etc. – to control the hazards. Only a chemical engineer can remove the hazards by changes in design.

2. Most chemical engineers will never use much of the knowledge they acquired as students; they may never have to design a distillation column, for example, or operate a furnace. But all chemical engineers, whether they work in production, in design or in research, will have to take decisions on loss prevention, will have to identify the hazards on a new or existing plant, decide how far to go in removing them and the most appropriate way of removing them, and see that action is taken. Young engineers need to be made aware that they are responsible, both morally and legally, for the protection of their company's employees and of people who live near the plant and that they are the custodians of their company's assets and reputation.

Universities which give no training in loss prevention are not preparing their students for the tasks they will have to undertake once they graduate.

3. It is sometimes argued that the job of an university is to train students in the principles of a subject and that applications should come later when the student enters industry. However, accounts of serious accidents of the past can be used to illustrate scientific principles and to show how accidents can be prevented by the application of basic knowledge.

Not every one would agree that the job of a university is to train students only in the principles of a subject. One academic writes, "An engineer is effective only through what he does, not through what he can do or knows" (Mathews et al., 1985); one of the things he needs to do, to be effective as an engineer, is to design and operate safely.

4. The physical scientist and engineer is distinguished from the non-scientist, the sociologist and the life-scientist by an ability to apply numerical methods to the solution of problems. Hazard analysis, the application of quantitative methods to safety problems, as well as being an important study in its own right, can be used to show that many problems which at first sight do not seem to lend themselves to numerical treatment can, in fact, be treated quantitatively. There is never enough money available to remove all hazards. How do we decide which to remove as a matter of priority, which to live with, at least for the time being.

Assuming loss prevention is included in the training of undergraduate chemical engineers, what subjects and methods should be included? I suggest:

1. *Hazop*. Hazard and operability studies (hazops), a systematic method of identifying hazards, are specifically mentioned in the Institution of Chemical Engineer's syllabus. I suggest that the best way of teaching them, to undergraduates or mature students, is to describe the technique briefly (taking, say, half an hour) and then let the students apply it to a line diagram.

2. *Hazard Analysis*: Systematic quantification of hazards is specifically mentioned in the I.Chem.E syllabus. Aspects that should be covered are the calculation of probabilities including fault trees, pitfalls, criteria and simple applications such as the estimation of test frequencies. The objective is not merely to introduce students to methods of calculation but to get them to realise that a systematic approach is possible to questions such as "How far should we go in removing hazards?" and "Is the cost of this measure justified by the size of the risk?"

3. *Inherently Safer and Friendly Design*: Although this is not specifically mentioned in the I.Chem.E syllabus it is such an important part of loss prevention and will become increasingly so during the lifetimes of today's students that it should be included in every syllabus. Students can be asked to apply the principles of inherently safer design to their design project.

4. *Human Error*: Although not specifically mentioned in the I.Chem.E syllabus this is an essential part of management for safety, which is included. Many industrial accidents are said to be due to human error. Students should be encouraged to consider the extent to which accidents can be prevented by persuading people to take more care, the extent to which we should try to change the work situation, that is, the design or method of operation.

5. *Maintenance and Modification*: These are specifically mentioned in the I.Chem.E syllabus and they can supply sets of notes and slides which can be used for the discussion of accidents that have occurred involving maintenance and modifications (and other subjects). The discussion leader describes the accident very briefly, the class question him to establish the rest of the facts and then say what *they think* ought to be done to prevent the accident happening again.

6. *The Law*: I have put this last because the primary responsibility of the engineer is to do what is necessary to protect people and property and this will usually involve doing more than the law requires. The principles of the law, rather than details of regulations, should be taught.

In the U.K. there is long tradition of giving practical training to scientists without compromising on the teaching of fundamentals. Reviewing a book on Victorian chemistry, Hamlin (1985) writes, "... The 'pure science' curriculum they developed turned out to match remarkably well the kinds of tasks that industrial chemists ended up doing; ... at the center of the curriculum was a saleable skill; ... Professors argued that chemistry was worthy because it combined practicality with mental discipline".

LEARNING FROM EXPERIENCE (Kletz, 1988b)

We learn more from mistakes than from successes but we cannot learn from mistakes unless we are told about them. Unfortunately companies are becoming increasingly reluctant to publish details of their accidents.

Before discussing the reasons it may be useful to say why we should spread information on accidents that have occurred.

1. The first reason is moral. If we have information that might prevent an accident then we have a duty to pass on that information to those concerned. If there is a hole in the road and we see someone about to fall into it, we have a duty to warn him. If we fail to do so we are, to some extent, responsible for his injury. The 'holes' into which oil and chemical plants fall are more complex than a hole in the road but the principle is the same.

2. The second reason is pragmatic. If we tell other people about our accidents, then in return they may tell us about theirs, and we shall be able to prevent them happening to us.

3. The third reason is economic. Many companies spend more on safety measures than some of their competitors and thus pay a sort of self-imposed tax. If we tell our competitors about the action we took after an accident, they may spend as much as we have done on preventing that accident happening again.

4. The fourth reason is that if one company has a serious accident, the whole industry suffers in loss of public esteem while new legislation may effect the whole industry. So far as the public and politicians are concerned, we are one. Flixborough cost the insurance companies about $100 million in material damage. It cost the rest of the chemical industry much more. The same is true of Seveso and Bhopal. To misquote the well-known words of the English poet, John Donne:

"No plant is an Island, entire of itself; every plant is a piece of the Continent, a part of the main. Any plant's loss diminishes us, because we are involved in the Industry: and therefore never send to know for whom the inquiry sitteth; it sitteth for thee."

Why are we circulating fewer accident reports?

1. The first and most important reason is time. Since the recession in 1980 many companies have reduced staff and those who are left have less time for putting accident reports into a form suitable for publication or attending conferences to present them. When people are under pressure, inevitably jobs that they intend to do and know they ought to do but do not have to do by a specific

time, get repeatedly postponed, especially when these jobs will bring them less credit with their employers than attending to output and efficiency.

2. The second reason is the influence of company lawyers, particularly in the U.S.A. where they are much more powerful than in the U.K. They often say that publication of an accident report or even circulation within the company may effect claims for compensation or lay the company open to prosecution.

3. Another reason is fear of adverse publicity. If an accident is described at a conference or in a technical journal, perhaps the press or TV will pick it up and use the incident to show that the company is careless, irresponsible and so on. This does not seem to happen in practice. If it did, the company could say that publication showed their sense of responsibility.

However, if fear of adverse publicity is not a valid reason at the company level, it is powerful at the individual or group level. People are naturally reluctant to draw attention to their failures and expose their weaknesses, particularly if the company culture is intolerant of error.

4. The procedures to be followed in many companies are so laborious that they discourage publication.

5. It is sometimes said that publication of accident reports may result in disclosure of confidential information. These fears are unfounded. Inessential details can be altered to avoid giving away secrets without destroying the essential safety message.

What can we do to encourage publication?

1. Encouragement from the top is of primary importance. Statements of policy count for little. A note from a senior manager saying "I liked your paper on the fire we had last year" or "Are you going to publish the report on the gas leak?" is much more effective.

2. Companies, and particularly their lawyers, are usually more willing to let one publish reports on near-misses, in which no one was injured, than reports on accidents that resulted in injury or death. We can learn just as much from these near-misses as from other accidents.

3. Anonymity helps. Accidents do not reflect credit on those concerned and no one likes to be pilloried in public. Accident reports, even for internal company circulation, should be edited so that the plant or factory where the accident occurred cannot be identified.

4. If we cannot publish the report on a accident perhaps we can publish details of the action we took as a result. This may not have the same impact as the report but it is a lot better than nothing. For example, in 1977 reports appeared in the technical press of a fire on a large refrigerated propane storage tank. The company concerned said that for legal reasons they could not publish a report on the incident and, so far as I am aware, no report has ever appeared. However an employee of the company did present several conference papers

describing new standards for cryogenic storage. It is not difficult to read between the lines and see what happened (Cupurus, 1979, 1980, 1981).

5. Finally, stand up to the company lawyer. Ask him if he really wants to let people fall into that hole in the road.

POLICIES

In theory the big men at the top lay down policy and the rest of us follow it. In practice, we deal with problems as well as we can, subject to various constraints. Looking back we see a common pattern. That is our policy. If we want to change a policy we should not start by asking the Board to issue a policy statement. Instead we should persuade individual managers, the lowest level that has the power to make the changes we want. Gradually more and more of them will do what we want and it will become the policy.

According to David Lodge (1988), the Englishman does not worry about big things such as policies, religion and death. (We have to use German words if we want to discuss the *Zeitgeist* or *Weltanschauung.*) He is "more empirical: he worries about whether he is on the right train, about how much to tip the taxi-driver; he gets up in the middle of the night to see if he turned off the living room light". Success in safety depends on this sort of worry. (Has the protective system been tested? Are there any gaps in the fire insulation? Are the instructions unambiguous?)

HUMAN ERROR

If this conference had been held two hundred years ago there would probably have been a paper on phlogiston. That theory has long been abandoned. It did not help us to forecast the behaviour of substances.

In a book on radio published fifty years ago there is a chapter on the ether. It says, "The whole of space is filled with ether – a continuous very dense and very elastic medium ... the ether itself consists of very fine particles which, though practically incompressible, can easily be shifted in respect to each other."

Today no one would write like this. We have realised that the concept of the ether is unnecessary. It does not explain anything that cannot be explained without it.

Similarly, biologists have abandoned the idea of protoplasm as the physical basis of life, something which permeates inanimate structures and gives them vitality.

I suggest that the time has come when the concept of human error ought to go the way of phlogiston, the ether and protoplasm. For several reasons it hinders rather than helps us in identifying the causes of accidents and the action required to prevent them happening again:

1. Any accident can be said to be due to human error. If someone, operator,

maintenance man, designer or manager had done something differently the accident would not have occurred. If equipment failed, the designer could have chosen better equipment. (However, when an accident is said to be due to human error the manager who completes the accident report usually means an error by an operator or maintenance man. Managers and designers are apparently not human, or do not make errors.)

2. 'Human error' groups together widely different phenomena which call for different actions. There is little in common between those human error accidents that occur because someone did not know what to do (preventable by better training or instructions), those that occur because someone decided not to do what they knew they should do, those that occur because someone lacked the physical or mental ability to do what they should do, and those that occur because someone has a moment's forgetfulness.

3. If we say an accident is due to human error then the next step, usually, is to tell someone to take more care. But that will not prevent any accidents. In contrast, if we ask what action is needed to prevent the accident happening again, people may suggest better training or instructions, better enforcement of instructions, or better plant designs or methods of working which will reduce or remove opportunities for error.

In short, let us stop using the term 'human error', stop worrying about causes, and instead concentrate on the actions required. Perhaps it is 'cause' and not just 'human error' that ought to go the way of phlogiston, ether and protoplasm.

REFERENCES

Cupurus, N.J., 1979. Cryogenic storage facilities for LNG and NGL. In: Proc. 10th World Petroleum Congress. Heyden, London, p. 119.

Cupurus, N.J., 1980. Developments in cryogenic storage tanks. Presented at the 6th International Conference on Liquefied Natural Gas, Kyoto, Japan.

Cupurus, N.J., 1981. Storage of LNG and NGL. Presented at the Seminar on LNG and NGL in Western Europe in the 1980s, Oslo, 2 April.

Hamlin, C., 1985. A vision of chemistry. CHOC (Center for History of Chemistry) News, p. 17.

Kletz, T.A., 1988a. Should undergraduates be instructed in loss prevention? Plant/Oper. Prog., 7(2): 95.

Kletz, T.A., 1988b. On the need to publish more case histories. Plant/Oper. Prog., 7(3): 145.

Lodge, D., 1988. Write On. Penguin Books, p. 5.

Mathews, A.T., Main, A.N. and Beveridge, G.S.G., 1985. Getting Experience. Engineering Design Education, Spring: 3.

CASE STUDIES: Industrial risk management in practice

Process safety assessment of new and existing plants*

Jim Pikaar, John Braithwaite and Tony Cox
Shell Internationale Petroleum Maatschappij B.V., The Hague, The Netherlands

ABSTRACT

Pikaar, M.J., Braithwaite, J.M. and Cox, A.P., 1990. Process safety assessment of new and existing plants. *Journal of Occupational Accidents,* 13: 103–110.

Companies of the Royal Dutch/Shell Group operate a very wide range of manufacturing installations which include oil refineries, gas processing and liquefaction plants and chemical plants.

The safety assessment of these installations is approached from a background of experience. The necessary experience to do this is continuously brought to bear in design by involving appropriately selected staff and by applying established guidelines, standards and design/engineering procedures throughout the development of a project. In addition, structured safety reviews are carried out to check that all aspects have been properly dealt with.

The safety assessment of existing plants follows a pattern of enquiry which is similar to that for new plants.

Finally the question of residual risk is addressed: in general the assessment of the consequences of a failure scenario is more reliable than the judgement of its probability. Nevertheless there are instances where a closer look at probabilities can help in making design decisions. It is questionable whether risk analysis results can be used meaningfully in an absolute sense to judge the acceptability of an installation.

INTRODUCTION

The safe operation of a process plant can only be achieved if the people involved throughout its life – from the initial process design concept to its final shutdown – approach their tasks with a high degree of safety awareness backed up by knowledge and experience. This applies equally to all staff whether they design, build or operate plants.

In this paper we shall deal with the way process engineers help to make plants safe. Though the contributions of members of other disciplines, such as

*Presented at the International Conference on Industrial Risk Management, Zürich, Switzerland, 16–17 January 1989 (first presented at the International Symposium on Loss Prevention and Safety Promotion in the Process Industry, Cannes, France, September 1986).

equipment specialists, plant operators and maintenance engineers to name but a few, may only be mentioned in passing, the quality of their work and their safety consciousness are of course also of vital importance to achieving a good safety record. The paper concentrates on the safety assessment of a plant as designed and built and does not cover matters such as plant management and how over the years a safety culture is built up in the teams that operate the plant.

Process engineers in specialist safety groups work together with process designers, advising on safeguarding aspects of the design during its development. They are also involved in auditing; they participate in the safety reviews which are held during the design and construction phases of a project and also on existing plants.

Such safety specialists provide advice in the area of hazard analysis, bringing available knowledge and experience to bear in judging accident scenarios and in assessing the consequences resulting from dispersion and/or combustion.

DEVELOPING AND RECORDING EXPERIENCE

The conviction that experience forms the mainstay of achieving safe plants is universally endorsed. In view of its paramount importance it is worth describing the forms it can take, how it is acquired and how it is maintained within a large group of companies operating in different countries.

Many Shell companies own, operate and maintain process plants based on designs which they acquired or which they themselves developed. Service companies, located in The Hague, may be called upon to provide services not only to design plants and make arrangements for their construction and commissioning, but also to advise and assist in each phase of the operating life of the plant. This interaction between engineers in the service companies and their colleagues in operating companies ensures familiarity with a wide variety of matters concerning the installations which form part of oil refineries, gas processing and liquefaction plants and chemical plants. Feedback on incidents which occur, and on the performance of safety systems and equipment, is made available to the designers of new plants and is also incorporated in the safety practices applied to existing plants.

Direct experience of the practices and problems of plant operation is fostered among the engineers in the service companies by frequent interchange of personnel with members of the operating and maintenance functions in the field. Such transfers are also of benefit to the operating companies as they provide them with technologists who have been involved in the latest designs and who bring with them a new awareness and an eye for possible improvements. This applies as much to safety aspects of the plants as to their efficiency.

Accidents involving injury or more than a certain value of material damage are reported to the service companies to enable the lessons learned to be shared.

A few years ago the reporting system was extended on a trial basis in a number of countries to cover hazardous events in which a release of flammable or toxic material occurred, but which did not necessarily lead to injury or consequential damage. The data bank of operating incidents being assembled in this way is considered valuable. Safety bulletins and newsletters are distributed widely in Shell companies to inform staff of these experiences, giving specific warnings and pointing to more general lessons that can be learned.

Safety training courses are organised by service companies and operating companies and these are attended by mixed groups of designers and operators. Subjects covered include the design of safety systems, the hazards of releases and their mitigation, and the safety features incorporated in various types of processes.

The experience and knowledge that is continually developed in Shell companies is recorded by the service companies in formal documents, such as Design and Engineering Practices, which form part of the recommended specification for new installations. There are also manuals for the design, engineering and operation of whole installations and guidelines with recommended approaches to commonly occurring situations in design and operation.

SAFETY IN NEW DESIGNS

Process safety reviews

It is important that the safeguarding measures applied in the design are documented so that they can be included in the plant design book and eventually in the operating manual. In the course of the design of a new plant, when process and instrument diagrams are more or less firm, it is now common practice to summarise all the process safeguarding measures in a Safeguarding Memorandum or a Safety Report.

A Safeguarding Memorandum includes:

- Review of the relief cases for each safety relief device, highlighting the determining case;
- Summary of the design intention of the instrument safeguarding systems;
- Description of the emergency shutdown and depressuring systems;
- Section on the interfaces between the process unit under examination and the adjacent process units, utilities and storage systems. Basically, this is to answer the question "What effect can the new unit have in terms of overpressure, overheating, contamination, etc. on other systems?"

This Memorandum has proved to be particularly useful in the design of high pressure oil refining and gas processing plants, where the consequences of a process breakthrough via a high pressure/low pressure interface can be severe.

An appendix to the Safeguarding Memorandum is the process safeguarding

flow scheme, a simplified version of the process and instrument diagrams, usually drawn as a single sheet for each process unit. The flow scheme shows only:
- Measures that constitute the last line of defence against loss of containment, e.g. relief valves and high temperature trips, but not high-pressure trips acting as pre-relief protection;
- Elements that influence the size of relief valves, e.g. certain control valves and bypasses;
- Measures intended to limit the extent of a possible loss of containment, e.g. remotely operated valves in pump suction lines, and emergency depressuring valves.

The process safeguarding flow scheme was devised as a means of highlighting to operating personnel the most important process safeguarding measures, and in particular items such as control valve sizes which, if altered, could adversely affect the protective measures.

The preparation of these documents – the Safeguarding Memorandum and the process safeguarding flow scheme – at the end of the process engineering phase of a design has also proved to be an effective means of review. In particular the single-sheet Flow Sheet facilitates the checking of the interfaces between the new unit and existing systems.

An investigation and recording procedure has been developed for chemical plants that covers the health, environment and safety aspects of their design. This is particularly important for plants handling materials that can present either long-term or short-term health hazards. The Safety Report associated with this procedure, in addition to overpressure protection, covers aspects of the design concerned with potential runaway reactions in batch plant, dust explosion hazards and other matters not necessarily relevant for the "mainstream" continuous petroleum and petrochemical processes.

As with the Safeguarding Memorandum the object of the Safety Report is to document all safety aspects raised during the design, so that they can be made available to the operating personnel.

At discrete stages in the development of a project a number of safety studies of identified content and scope are carried out.

Process safety reviews

Often, a first process safety study is carried out during the feasibility or scouting stage of a project. This is when a process has been defined in terms of a simplified flow scheme, and a site has been identified. The study concentrates on the potential hazards of the materials to be used in the process and possible harmful effects of the plant on its surroundings.

The first structured process safety review is normally conducted at the end of the process design phase of a project when the process and instrument diagrams are available. A further review is carried out during the detailed engineering (implementation) phase of a project by which time the process and instrument diagrams have been frozen and most equipment vendor data are available. These reviews are undertaken by multidisciplinary teams including process designer, project engineer, safety specialists, and, preferably, a representative from the future operating team in order to ensure continuity through to start-up and the running unit. Normally, the team members have not been closely involved so the design can be scrutinised by fresh eyes.

The objectives of these reviews are to ensure that applicable codes of practice have been followed, current safeguarding practice has been incorporated and experience from similar running units has been built into the design.

The team works systematically through the process and instrument diagrams and supporting documentation, critically examining the equipment arrangements and the safeguarding features and highlighting any weaknesses in the design. As the examination is carried out in meticulous detail and is therefore rather demanding, sessions are normally limited to half days. The remainder of the time is made available for individual team members to investigate further items and queries raised during the examination sessions.

HAZOP studies

The formal HAZOP (hazard and operability studies) procedure is applied to sections or whole plant where there is considered to be a very significant potential for hazard, on account of large inventories of flammable or toxic material, or because of extremes of process conditions. It is also applied where a process has undergone a significant development and the extrapolation of knowledge from existing units does not provide sufficient experience of safety aspects.

It is our experience that the application of the methods systematically to all sections of a plant design, apart from being very time consuming, can lead to a dilution of the creative approach which is a key element in the method.

Piping models

Piping models – scaled down plant models, complete with piping – are normally constructed during the detailed engineering phase of new designs and major revamps. They were originally intended as an aid to engineering design and construction, and traditionally provided a means of checking the ease of access and escape. Now, they are increasingly being used to review in detail three-dimensional aspects of process safety and operability that cannot be picked out from process and instrument diagrams. Examples of aspects that can be checked on the piping model are:

- Safe location of vents with respect to possible flammable and toxic vapour release;

– Avoidance of low points and dead ends in piping systems; and
– Location of gas detection and other safety and fire fighting equipment.

Site visits

A site visit just prior to start-up of a new plant is recommended; it gives the opportunity to follow up queries raised during the process and instrument diagram reviews that cannot be answered by referring to design documentation, as well as to check on operational and personal safety measures incorporated in the plant as built.

PROCESS SAFETY OF EXISTING PLANTS

Once a plant has been built and is running, its safety aspects continue to be reviewed by the site. This is stimulated by the exchange of information and personnel described in the section on developing and recording experience.

In addition, the safety of existing units is also reviewed by the Operating Company in a more systematic manner following the same general principles as for new units. Of prime importance before starting such a review is to ensure that the process and instrument diagrams and supporting design information are up to date. These reviews also comprise a process and instrument diagram study and site inspection, and particular emphasis is placed on identifying the areas where codes and safeguarding practice have become more stringent since the plant was built.

The preparation of a Safeguarding Memorandum with a process safeguarding flow scheme is also proving to be an effective means of reviewing the safeguarding of older plants and identifying areas in which they differ from more recent designs. In assessing the desirability of an operational or design change the site may seek advice from the service companies.

THE RESIDUAL RISK

However efficient safety evaluations may be, there will always be a residual risk that some chain of events or unforeseen situation might occur and result in a major release of flammable or toxic material to atmosphere. The exchange of information at loss prevention symposia and other activities, such as inter-industry working groups on design standards, are ways by which industry seeks to reduce residual risk.

Shell companies have for some time recognised the need to be able to answer questions about possible major accidents, particularly when there is a step forward in technology or an increase in plant size. An example of the latter has been the need to find ways of assessing what might happen if large quantities of liquefied natural gas or other hydrocarbons were to be spilt. The considerable amount of research work done by Shell laboratories in the fields of heavy gas dispersion, radiation emission from fires and other areas associated with releases of hydrocarbons has results in papers to loss prevention symposia. As a result of this work, safety specialists of Shell companies have available an extremely valuable set of computer programs for consequence calculations.

These consequence models have been used to assess the possible effects that releases from plants may have on the site and the surrounding population. The results may lead to a closer review of plant design and method of operation which in turn may prompt engineers to study how such a release can be prevented, or its probability reduced so that its effects can be minimised.

In a recent project, the possibility of unacceptable concentrations of toxic gas (hydrogen sulphide) outside the plant area was identified for certain release scenarios. As this was highlighted at an early stage in the project, further studies into improved process design, better layout and higher engineering standards enabled changes to be incorporated into the design.

A very large amount of time can be spent on quantifying the worst possible effect of a large release of a flammable or toxic gas without taking into account either the probability that it will occur, or the likelihood that it will actually damage property or hurt people. This "maximum credible accident" approach can be useful at the early stages of a plant design, when possible hazards are being evaluated. However, it is important to maintain a proper perspective in the light of experience when judging the results of such evaluations. Due weight should be given to the greater relative likelihood of smaller accidents in order that safety measures are adopted that are most effective in terms of overall safety.

The quantification of risk calls for estimates of both physical consequences and the probability of postulated events. Each of these estimates contains a degree of uncertainty. The assessment of consequences of an event is based on the scientific understanding of physical and chemical processes; uncertainty here is due to applying simplified models to what are complex real situations. This is different to the uncertainty inherent in probability estimates for unlikely events, where a considerable amount of speculation may be involved, yielding numbers of doubtful value. For this reason quantification of probabilities does not play a part formally in the overall safety strategy of the manufacturing functions of Shell companies. However, it is used selectively by safety auditing groups to assist plant designers and operators to make decisions about safety problems. It can only be used in situations where failure rate data exist or can be judged realistically. The figures are then used to compare design alternatives, or to compare a proposal with the existing situation. In doing this, one has to bear in mind the uncertainty inherent in the figures used for the analysis.

Typical situations where the techniques have been used are:

- Review of instrument safety systems;

- Consideration of the risks associated with extreme weather conditions in a typhoon area;
- Choice of scenarios for safety studies on pressurised LPG storage installations;
- Review of the safety of a refrigerated propylene storage installation; and
- As part of safety studies of the transport of products by road, rail and pipeline.

Overall quantification of risk finds little use in the manufacturing functions of Shell companies. We are uncertain how much its use will grow, or needs to grow. It is recognised that most of the judgements of risk that form part of safety decision making will continue to be made on a qualitative basis.

The development of the techniques within the industry is watched and their potential is monitored. As an example, several years ago a licence was acquired for Shell companies to use a commercial computer package for performing risk analysis calculations. The package incorporates consequence models and can produce individual risk and societal risk results. When used by an experienced analyst there are a limited number of situations where this powerful tool can help compare the safety of alternative designs. However, we do not believe that, in the present stage of the development of risk analysis, this or any other method of risk analysis computation gives values of risk that can be used meaningfully in an absolute sense for making decisions about the residual risk associated with process plants.

In the final analysis many of the major accidents in the industry have occurred because of mistakes or errors in judgement made by the people involved in the day-to-day running of the plant. In order to keep the overall residual risk at a low level, there must be a continuous effort to ensure that the experience gained in the design and in the operation of plants is maintained. Associated with this, the systems for reporting incidents and recording the reasons for design decisions must continue and must be part of ongoing safety training programmes for both those who manage and those who operate the plants.

Workplace analysis and design solutions in the Dutch rubber industry*

P.H.J.J. Swuste

Safety Science Group, Delft University of Technology, P.O. Box 5050, 2600 GB Delft, The Netherlands

ABSTRACT

Swuste, P.H.J.J., 1990. Workplace analysis and design solutions in the Dutch rubber industry. *Journal of Occupational Accidents,* 13: 111–120.

A method of solution directed workplace analysis is developed to compared and evaluate different production processes within the branch of the Dutch rubber industry as well as their solutions to reduce exposure to different factors. Special attention is paid to one preventive measure; local exhaust ventilation. Efficiency and design features will be discussed.

INTRODUCTION

The rubber industry has a long history of health and safety studies. Most of these studies were conducted by epidemiologists or toxicologists. The bladder cancer story in this sector of industry is probably well known to everybody (Case and Hosker, 1954). In the last century and the beginning of this century health risks in the rubber industry were reported by hygienists in France, Germany and Great Britain (Delpech, 1863; Arlidge, 1892; Weyl, 1897; Oliver, 1902; Collis and Greenwood, 1921). Apart from accidents at mixing mills, much attention was given to the health effects of exposure to 'naphtha' and CS_2 (carbon disulfide). Both chemicals were used extensively in the rubber industry. Naphtha as a solvent and CS_2 as a curing agent. In particular the long term effects of CS_2 were described and were considered to be as serious as the famous 'hystery saturnine', lead intoxication, and 'phossy jaw', phosphorus necrosis. Some authors paid particular attention to preventive measures, with dilution and local exhaust ventilation as the most important ones (Weyl, 1897; Oliver, 1902). From 1920 onwards the risk of bladder cancer was mentioned, as a result of exposure to the accelerator aniline, and of skin and scrotum cancer due to exposures to naphtha derivatives (Heyermans, 1926). More recently cancer risks of the lungs, the gastrointestinal tract and asbestos related mesothelioma have been reported (Fox et al., 1974; McMichael et al., 1976; Baxter and Werner, 1980; Sorahan et al., 1986). Also increased respiratory morbidity and reduced pulmonary function is described in studies among rubber processing workers (Fine and Peters, 1976). And the skin as a possibly underrated route of intake has been stressed (Kromhout et al., 1988).

Compared with studies to health effects, studies on exposure levels to different factors are more scarce. This applies even more to studies of the effectiveness of different types of measures to reduce exposure.

In a study carried out by the Safety Science Group of the Delft University of Technology and the Departments of Air Pollution and Environmental Health of the Agricultural University of Wageningen and evaluation has been made of the exposure levels and measures to prevent exposure in the Dutch rubber industry. Ten rubber companies out of the total of 48 Dutch rubber firms were selected. This selection is representative of this branch of industry and covers a large range of firms on criteria like number of employees and nature of production. The companies also were representative in respect of membership of an occupational health service, presence of shop stewards and of works councils.

In the companies the exposure to chemicals, noise and physical stress was investigated by means of personal and environmental monitoring. This quantitative approach enabled assessments to be made of the health risks and circumstances responsible for the level of exposure, as well as of the quality of existing measures to reduce exposure. This last item in particular gives the opportunity to compare and evaluate preventive measures. It has been the experience of the research group in Delft, that many successful solutions to exposure problems, developed in one company, were never properly evaluated (Musson et al., 1988). And, furthermore, there was little or no cross-fertilisation of ideas, whereby solutions were passed from one company to another. Therefore the project was used to develop a method for a solution-directed workplace analysis and evaluation.

This paper aims to give the main characteristics of this method with particular attention to the production process and its organisation. As a consequence in this paper the subjects of personal measurements and assessment of health risk receives less attention.

SOLUTION-DIRECTED WORKPLACE ANALYSIS

The method starts with a functional analysis of the production process, making use of a threefold classification developed in the field of system design. The three levels are interrelated and organised hierarchical. It can be used to describe and compare different production processes.

Production function. The production function is an element in the production process. It defines the function or activity to be carried out. For example weighing of chemicals (compounding) or curing.

Production principle. This defines the different principles by which the function can be carried out. For example, these can be manual, mechanical, remote controlled or automated. Curing of rubber products can be a remote controlled continuous process e.g. infrared, salt-bath or ultra high frequency vulcanisation. Another production principle of curing is mechanically served injection or compression moulding.

Method of execution or design or form level. The form level is the actual design of the production principle. It describes the machine used and the measures to prevent accidents or exposure, e.g., Lunn bars or other emergency-stop mechanisms and local exhaust ventilation.

The production process in rubber companies can be divided into seven production functions (Table 1). As an example of the analysis, the production function 'weighing of chemicals – compounding' is pictured in Table 2.

TABLE 1

Production functions in rubber production processes

Production functions in rubber production processes
Supply of raw materials
Weighing of chemicals
Mixing raw materials
Pretreating unvulcanised rubber products
Modelling unvulcanised rubber products
Curing
Finishing

TABLE 2

Production principles and methods of execution of the production function 'weighing of chemicals – compounding'

Production function	Weighing of chemicals – compounding			
Production principle	Manual	Mechanical	Remote controlled	Automatic
Design	Filling station Empty bag press Open bags Store bins Grocers scoop Weighing scale Plastic or paper bags	Semi-automatic weighing system		Automated weighing and batching system

PREVENTION STRATEGY

The functional analysis, pictured in Tables 1 and 2 is related to different forms of exposure. Two prevention strategies are distinguished:

Elimination of the source. The source of exposure is no longer used.

Control of exposure. A barrier is established between the source and the worker.

Table 3 pictures the relation between the prevention strategies and the functional analysis. Elimination of exposure is achieved by reorganising a production process, like combining or erasing production functions. So the production function compounding, prior to mixing becomes redundant with the introduction of premixed rubbers. Introduction of an automatic supply of chemicals to the mixing function is an example of elimination by changing production principles.

Control of exposure is the second prevention strategy. Adopting local exhaust ventilation on machines is such a prevention measure on the level of the method of execution. After this division the different production principles, designs and preventive measures can be compared and evaluated both qualitatively and quantitatively. The comparison gives a view on potential solutions which are available in practice.

A summary of the most important prevention measure for the production functions 'weighing of chemicals – compounding', 'mixing raw materials' and 'curing rubber products' is given in Table 4. Some examples of preventive measures of Table 4 will be discussed in the next paragraph.

RESULTS

In this study the evaluation of preventive measures was restricted to local exhaust ventilation. This type of prevention, through control of exposure, is one of the most frequently applied measures in Dutch rubber companies. In that respect the rubber industry is not unique.

Local exhaust ventilation was analysed in our study both qualitatively and

TABLE 3

Possibilities of prevention. Relation between prevention strategies and functional analysis

	Prevention strategies	
	Elimination	Control
Production function	×	
Production principle	×	×
Design		×

TABLE 4

Prevention measures for the production functions 'weighing of chemicals – compounding', 'mixing raw materials' and 'curing rubber products' in ten Dutch rubber firms

	Preventive measures by compounding and mixing	
	Elimination	Control
Production function	Dust free forms of chemicals C-overloaded compounds Premixed compounds	
Production principle	Automatic weighing	Semi-automatic weighing system
Design		Plastic weighing bags Lunn bags Emergency-stop **Empty bag press** **Internal mixer** **Local exhaust ventilation**
	Preventive measures by curing	
	Elimination	Control
Production function		
Production principle		Injection moulding press
Design		Automatic press-unloading Enclosed injection press **Continuous vulcanisation** **Finishing table** **Local exhaust ventilation**

quantitatively. The qualitative analysis focussed primary on a summary of the methods of execution, provided with local exhaust ventilation. Besides presence, local exhaust ventilation was evaluated on the criteria of construction, effectiveness and maintenance.

Four main types of ventilation were distinguished on the criterium construction: enclosures, booths, captor hoods and receptor hoods. A judgement on this criterium was based upon observation and comparison of the local exhaust ventilation used with that required. The required type of ventilation was defined as that expected in view of the nature of the exposure source.

A judgement of the effectiveness was based on a number of observations.

First the possibility of disturbance of the direction of ventilation. Secondly the position of the employees' breathing zone during average work conditions in relation to the ventilation. Thirdly possible loss factors in the construction of the ventilation system. At last a judgement on this criterium is determined by smoke-tube test results.

Information on the frequency of maintenance and presence of openings for maintenance in the ventilation system was the criterium for the judgement on the quality maintenance.

For the quantitative analysis stationary dust measurements were carried out for all designs provided with local exhaust ventilation. We used a strategy of measurements developed and applied in an American study on dust and fume exposures in tyre manufacturing plants (McKinnery et al., 1984; Heitbrink and Crouse, 1984). Repeated dust measurements were taken in the direct vicinity and at a distance from the respective ventilation system. In this way both a source and background concentration were obtained. If the local exhaust ventilation system is working ideally both concentrations should be low. No difference in measurements indicates that the local exhaust ventilation is effective. A significant difference between the concentrations measured suggests that the emission source is not being controlled adequately. When the background measurements are significantly higher than source, it means that another source of emission is present in the same area. Because the experimental data has considerable variability, the significance of the concentration difference was evaluated using Wilcoxon's signed rank test.

Local exhaust ventilation in the production function 'weighing of chemicals – compounding' and 'mixing of raw materials'

From the production function 'weighing of chemicals – compounding' and mixing raw materials four important different detailed designs are presented in Fig. 1. This figure gives the total number of those designs found in the companies and those equipped with a local exhaust ventilation system.

In mixing departments two different production principles were found; the internal mixer and the open mixing mill (Fig. 1, resp. mixers and mills). The internal mixer is a remote controlled machine. The materials to be mixed are charged into a hopper and then into a chamber which contains the mixing rotors. The rotors mix the materials into a homogenous mixture. A conveyer belt is mostly used to charge the mixer. The open mixing mill is mechanically driven. The mill is charged manually by adding the chemicals and rubber between the two mill rolls. The rubber shapes itself in a continuous blanket around the mill roll and must be cut loose and fed around the roll a couple of times. This also is done manually. All charging doors of the internal mixers were provided with local exhaust ventilation. For mills it is only 50%. By reason of the large degree of enclosure of the source the ventilation of the mixers was

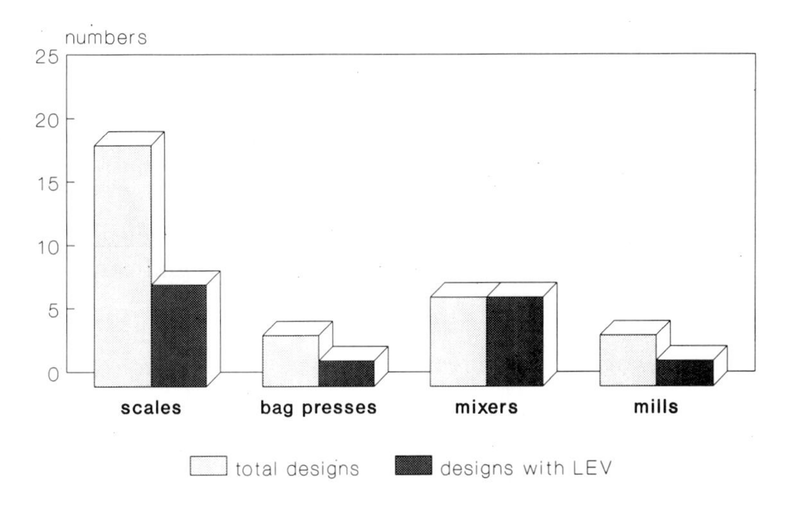

Fig. 1. Presence of local exhaust ventilation at the production function 'compounding'.

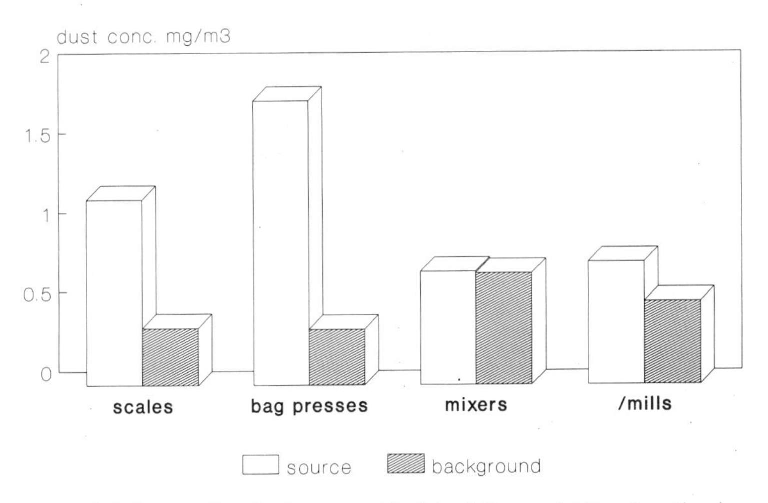

Fig. 2. Efficiency of local exhaust ventilation at the production function 'compounding'.

judged positively on construction and effectiveness in all cases. This is confirmed by the quantitative analysis, which gives no significant difference between source and background concentrations (Fig. 2). The open mixing mills were supplied with receptor hoods. This type of ventilation is effective for hot, gaseous components, but inefficient for dusts. The quantitative analysis too shows a difference between source and background measurements, which is significant.

Handling empty bags of powdered chemicals is an important source of dust exposure. Dust control measures are often overlooked. In only four cases were presses for empty bags found, two of them equipped with local exhaust ventilation (Fig. 1, bag presses). As with the ventilation by the mouth of the internal mixers, the ventilation of the bag presses also has a large degree of enclosure of the source. However the activity of the workers by the presses creates more dust. Bags of chemicals are cut open with a knife and emptied. For that reason the construction was judged positively and the effectiveness negatively. No information was available on maintenance. In the quantitative analysis the source and background concentrations differ with a level of significance of 10%.

Most rubber factories had more than one scale for weighing of chemicals. Less than 50% were ventilated (Fig. 1). The poor enclosure of the source and the large distance between source and hood made this ventilation less effective. Frequently the ventilation by the scales was suspended flexibly and in practice was turned away to give the least annoyance to the worker. The maintenance was also poor. The difference in concentrations was significant at 10%.

Local exhaust ventilation in the production function 'curing'

Use of remote controlled continuous curing principles is of a recent data. All the observed designs – ultra high frequency curing (uhf), salt-baths and hot air curing (cv) – had local ventilation systems (Fig. 3 resp. uhf, salt-bath, cv). They all function as a closed system. They were all judged positively on construction and effectiveness but not on maintenance. The same conclusion can be drawn from the measurements. Only the cv-curing had a high background concentration. This was caused by a powder bin in the same room, which created a high exposure to talc (Fig. 4, resp. uhf, salt-bath, cv).

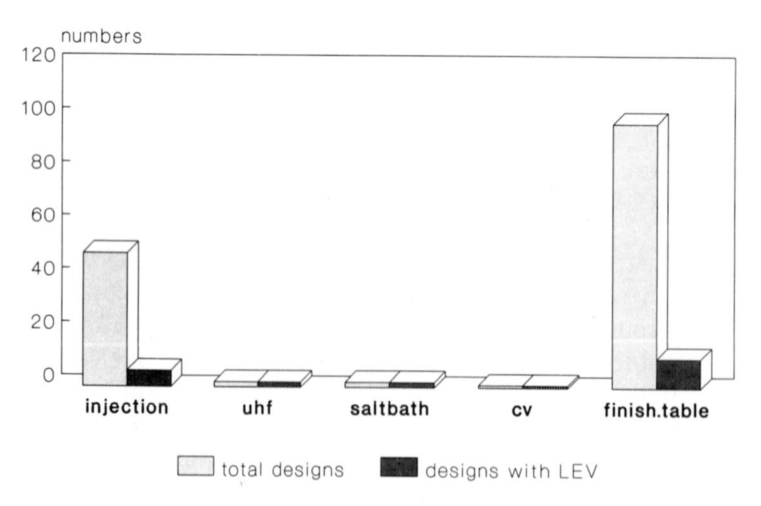

Fig. 3. Presence of local exhaust ventilation at the production function 'curing'.

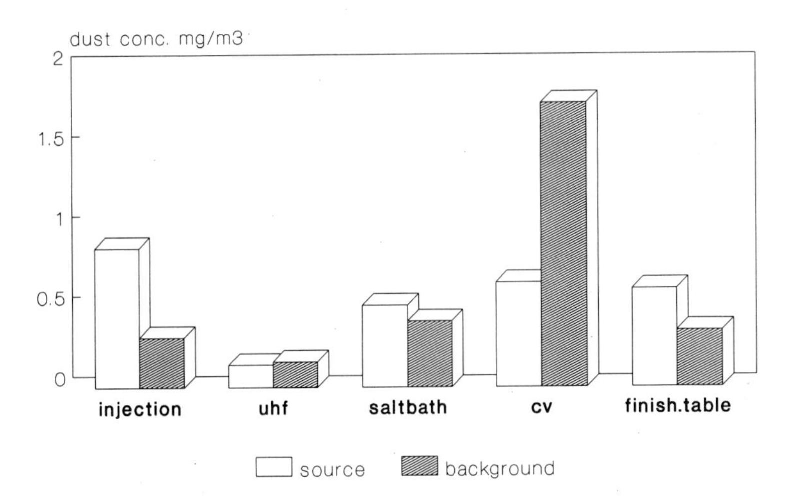

Fig. 4. Efficiency of local exhaust ventilation at the production function 'curing'.

The batch-type of vulcanisation is rarely equipped with ventilation. This applies even more for finishing tables by curing presses (Fig. 3, resp. injection and finish. tab). The construction of the ventilation was mainly judged as positive. The effectiveness and maintenance were worse. In practice the hot products were either put outside the reach of the ventilation or the finishing table was equipped with a downstream captor hood, while curing fumes from piled hollow products tend to rise. The quantitative analysis confirmed these findings.

CONCLUSIONS

The techniques developed in this study, for solution directed workplace analysis offer a systematic approach to the design of a production process. Within one branch of industry different types of production processes as well as solutions for the reduction of exposure can be compared.

With the exception of the use of pre-mixed rubber and dust-free forms of chemicals, most preventive measures in the Dutch rubber industry are focussed on control of exposure. This paper deals only with an assessment of the local exhaust ventilation. In the Dutch rubber industry local exhaust ventilation is mainly used at the start of the production process and more recently, where the principles of continuous curing is applied. The measure is effective in such continuous curing machines and also by the mouth of internal mixers. On the other hand handling of empty bags and the process of cooling cured products are hardly ever supplied with measures to reduce exposure.

Both the qualitative and the quantitative analysis confirm this picture. By the construction of local exhaust ventilation too little attention is paid to the nature of the source. This reduces the efficiency of the measure. This effect is made greater by the too small enclosure of the source and by the too great distance to the source.

Local exhaust ventilation is a preventive measure which is most used. It can contribute to dust and fume control, provided that its construction and maintenance are adequate and information of working conditions is considered in the draft of the ventilation system. The rubber companies do not, however, meet these conditions. This makes local exhaust ventilation insufficient where it is the only measure to reduce dust and fume exposure.

REFERENCES

Arlidge, I., 1892. The Hygiene, Diseases and Mortality of Occupations. Percival, London.

Baxter, P. and Werner, J., 1980. Mortality in the British rubber industries 1967-1976. HMSO, London.

Case, R. and Hosker, M., 1954. Tumor of the urinary bladder as an occupational disease in the rubber industry in England and Wales. Br. J. Prev. Soc. Med., 8: 39-50.

Collis, E. and Greenwood, M., 1921. The Health of the Industrial Worker. Churchill, London.

Delpech, M., 1863. Industrie du caoutchouc soufflé. Recherces sur l'intoxication spécial que détermine le sulfure de carbone. Ann. d'Hyg. Publ., 19: 65-183.

Fine, L. and Peters, J., 1976. Studies of respiratory morbidity in rubber workers.: III. Respiratory morbidity in processing workers. Arch. Environ. Health, 31: 136-140.

Fox, A., Lindars, D. and Owen, R., 1974. A survey of occupational cancer in the rubber and cable making industries: result of five year analysis, 1976-1971. Br. J. Ind. Med., 31: 140-151.

Heitbrink, W. and Crouse, W., 1984. Application of industrial hygiene air sampling data to the evaluation of controls for air contaminants. Amer. Ind. Hyg. Assoc. J., 45: 773-777.

Heyermans, H., 1962. Handleiding tot de Kennis der Beroepsziekten. Brusse, Rotterdam.

Kromhout, H., Ikink, H., de Haas W. and Bos, R., 1988. The relevance of the cyclohexane soluble fraction of rubber dusts and fumes for epidemiological research in the rubber industry. In: C. Hogstedt and C. Reuterwall (Eds.), Progress in Occupational Epidemiology. Excerpta Medica, Amsterdam.

McKinnery, W. and Heitbrink, W., 1984. Control of air contaminants in tire manufacturing. Research Report PB 85-173573. U.S. Department of Health and Human Service, Public Health Service, Center of Disease Control, National Institute for Occupational Safety and Health. NIOSH, Cincinnati.

McMichael, J., Andjelkovic, D. and Tyroler, H., 1976. Cancer mortality among rubber workers: an epidemiologic study. Ann. N.Y. Acad. Sci., 217: 125-137.

Musson, Y., Hoefnagels, W., Bakkeren, M., van Drimmelen, D. and Burdorf, A., 1988. Solution directed workplace analysis: removing mortar from brick walls. Paper presented at the Human Response to Vibration Meeting, 26-28 September 1988, INRS, Vandoeuvre.

Oliver, T., 1902. Dangerous Trades. The historical, social and legal aspects of industrial occupations as affecting health, by a number of experts. Murray, London.

Sorahan, T., Pares, H., Veys, C. and Waterhouse, J., 1986. Cancer mortality in the British rubber industry: 1948-1980. Br. J. Ind. Med., 43: 363-373.

Weyl, Th., 1897. Handbuch der Hygiene. Fischer, Jena.

Risk management for water and energy pipelines*

R.B. Kulkarni and A.S. Patwardhan

Woodward-Clyde Consultants, 500 12th Street, Oakland, CA 94607, U.S.A.

1. INTRODUCTION

Pipelines are linear systems usually serving a vital community function such as transport of water, fuel or industrial products. In case of utility pipelines a necessary design requirement is the maintenance of serviceability. Disruptions should be kept to minimum and safe operation is expected to be maintained at minimum maintenance effort. In case of energy pipelines such as oil or gas it can also be a public health and safety issue. Certain materials such as cast iron are susceptible to sudden breaks. Occasionally these breaks present a public hazard when escaping gas enters a building and causes an explosion. For example, the frequency of incidents reported to the Office of Pipeline Safety involving cast iron pipe is 1.5 incidents per 1,000 miles versus 0.7 for plastic and 0.5 for steel (Gideon and Smith, 1979). Also, the maintenance costs per mile of cast iron pipe are significantly higher than for steel or plastic pipe.

In this paper, we present a methodology which uses economic and risk exposure evaluations to make repair versus replace decisions, i.e., should a particular pipe segment be replaced in the planning year, or should it be maintained and repaired, as necessary, for at least one more year? The methodology provides an efficient screening tool to rapidly analyze an entire network of piping segments and identify the critical segments for replacement during the planning year based on probabilities of breaks and leaks, replacement and repair costs, and possible consequences of pipe failure. Although most of the

*Presented at the International Conference on Industrial Risk Management, Zürich, Switzerland, 16–17 January 1989.

discussion in the following sections pertains to gas pipelines it is also applicable to water pipelines.

In the past, the decision regarding whether to repair, renew, or replace a pipe segment has been generally made using personal judgement and historical leak data. Some attempts have been made to formalize the procedures for identifying leak-prone areas. These methods have utilized such statistical procedures as discriminant and regression analyses (e.g., Midwest Research Institute, 1982; Peters, 1983). Some of the past statistical studies failed to recognize the wide variation in operational characteristics and maintenance practices among utility companies. Furthermore, the accuracy of predictions was not verified in most of the studies. Under any case, statistical analyses by themselves do not provide any guidance to the decision maker as to the most cost-effective approach for dealing with a leak-prone pipeline system. A need, therefore, clearly existed for developing a formal decision tool that would allow the owners to utilize leak and other pipe-related data already in its file and to find the optimum approach to deal with potential problems in the system.

The desired characteristics of the methodology to be developed are described in the following section. The methodology that was actually developed is presented next. The remaining sections describe the results of testing and validating the methodology for two utilities.

2. DESIRED CHARACTERISTICS OF THE METHODOLOGY

The methodology should be capable of deciding which option is more economical: (a) to replace a given segment of a gas pipe at present time, or (b) to maintain it at least for one more year with repairs as needed during the year, and then decide whether or not to replace. In addition, the methodology should determine the priority order in which the critical segments should be replaced for any designated replacement budget. The evaluation of repair-versus-replace options should be based on a sound economic analysis of all relevant costs including:

- Replacement cost,
- Cost of leak and break repairs,
- Cost of operation, inspection, and maintenance,
- Cost savings (or penalties) associated with public work opportunities,
- Cost of repairing pending leak orders, and
- Cost of incidents.

In order to estimate future repair costs if a segment is not replaced, the methodology must be able to forecast future performance of the segment. Obviously, the exact times and locations of future leaks and breaks cannot be predicted. However, the methodology should provide statistical forecasts (i.e., estimates of probabilities of breaks and leaks) based on an analysis of past performance. The data used in the statistical analysis should be company-specific data (and not industry-average data) since maintenance practices and frequencies of failure do vary significantly from one company to another.

A high degree of confidence can be placed in the results of the methodology only if the accuracy of the estimated probabilities of breaks and leaks can be validated. The validation of the probability predictions should be based on independent data sets which were not used in developing the prediction models. A rigorous validation approach will be to compare predicted and observed frequencies if breaks and leaks on categories of segments with different levels of predicted hazard. For example, if the methodology predicts a relatively high (or low) probability of a break for certain segments, the observed frequency of breaks for these segments should also be high (or low).

Pipeline maintenance is a continuous dynamic process, with the action taken in any time period influencing conditions, decisions, and costs in subsequent periods. The decision methodology must address the dynamic nature of the maintenance problem and the planning process. Thus, to decide which action to take at the present time, one must analyze all possible future maintenance actions and consequences of these actions.

A methodology was developed which meets the desired characteristics discussed above. The next section describes the development of the methodology, its main models and their input–output characteristics, and the implementation of the methodology in the form of an integrated software system.

3. DEVELOPMENT OF THE MAINTENANCE OPTIMIZATION METHODOLOGY

It is convenient to divide a pipeline system into a network of segments so that repair versus replace options can be evaluated independently for each segment. We define a segment as a portion of a continuous pipeline with the same diameter and operating pressure, and with homogeneous characteristics with regard to design, construction, and past performance. Two decision models are described in this section – one to evaluate repair-versus-replace options for individual segments and the other to develop replacement priorities for the entire network.

A segmental decision model for evaluating maintenance options

Two major options need to be evaluated for a given segment each year:

1. Replace (or renew) the segment with the same or a different material, or
2. Maintain the existing segment for at least one more year, repairing it as necessary during the year, and then decide whether or not to replace.

If replaced, a sudden break on the new pipe would be unlikely however, a small annual cast will be incurred to operate and maintain the replaced pipe. If not replaced now, the probability of a break would depend on such pipeline characteristics as diameter, soil type, and leak history. In case of energy pipelines, if a break does occur, there is the possibility of gas migrating into a building and, in extreme cases, causing an explosion. In the case of an explosion, the agency policy would be to replace the pipe segment where a break occurred. It is also possible that gas from leaking joints will migrate into buildings. Although the probability of this is very small, the model considers it in its evaluation. If the segment is not replaced now, one would have to repair any joint leaks or breaks that may occur during the year. At the end of the planning year, the same tow options – replacing immediately or maintaining for at least one more year – would be available.

The sequential nature of the decision-making process can be represented in the form of a decision tree as shown in Fig. 1. The "tree" consists of two types of nodes (a decision node and a chance node) and several branches at each node. The branches at a decision node represent alternative actions available to the decision maker who will select one of these actions. At a chance node, the branches represent alternative consequences of the actions taken previously. The decision-maker cannot determine with certainty which consequence will actually take place. However, the probabilities of the various consequences can be estimated as a function of the relevant characteristics of the pipe segment.

For decisions involving uncertain consequences, minimizing the total expected costs is the appropriate criterion to choose among alternative actions.

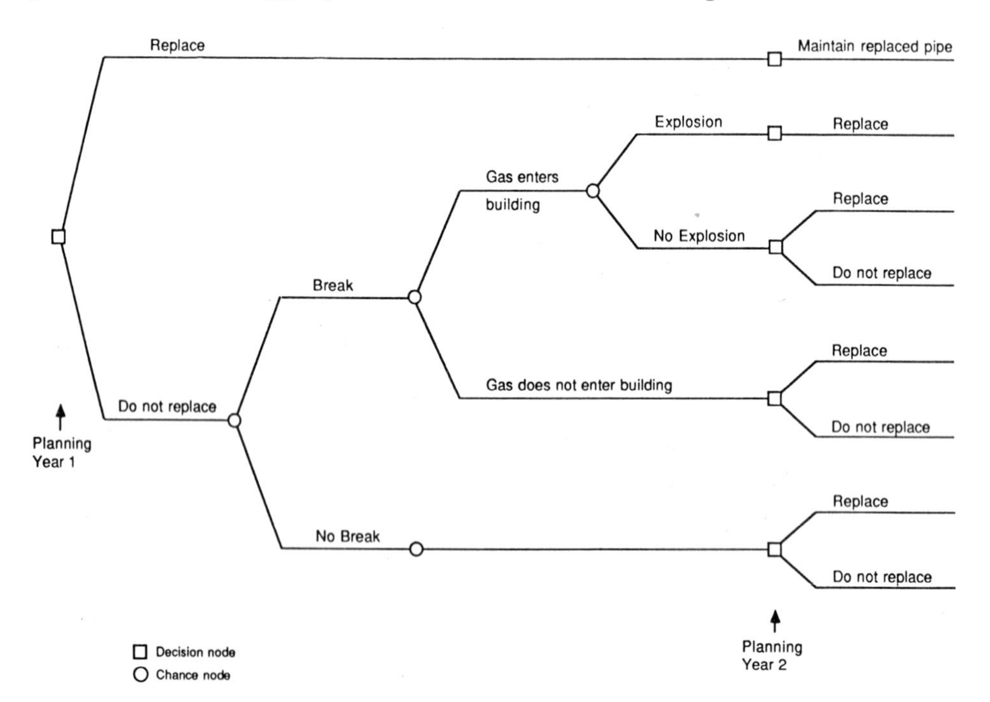

Fig. 1. Sequential replacement decisions under uncertain consequences.

Because of the sequential nature of decisions, one must analyze possible future consequences and actions that may be taken at future time periods before the choice of the best action at present time can be made.

The problem is formulated as a probabilistic dynamic model. It compares the cost of replacement (and maintaining the replaced pipe) to the total discounted expected costs associated with not replacing now but taking the optimal action next year. The costs associated with not replacing now are repair costs of expected and pending leaks, repair costs of potential breaks, cost of leaks, premiums associated with emergency repairs, penalties paid (sometimes in the form of more costly repairs) for digging up pavements which have been recently repaired, and inspection/preventive maintenance costs. The model also considers the opportunities of coordination with public work agencies allowing cost savings in street paving when replacing a pipeline and repairing pending leak orders. To calculate the replacements costs, the model considers the details of replacement procedure at the utility company including insertion and direct burial, type and size of piping material used, surface material and location of the replacement.

The consequence model that evaluates the expected costs associated with breaks (or leaks if gas enters buildings) considers actual as well as "adverse publicity" costs. It also has a provision for identifying "high consequence" segments such as those adjacent to schools and hospitals. These segments can be assigned higher consequence costs to give them a higher priority in replacement.

To solve the stochastic dynamic decision model, a threshold value is found for every segment that gives the optimal strategy. If the probability of break is larger than the threshold value the segment should be replaced; otherwise the segment should not be replaced this year. The threshold value is a function of all the costs and penalties described earlier, as well as the probabilities of gas entering buildings from leaks and breaks, and incidents if gas enters buildings. The threshold value also depends on whether "public work opportunity" exists for the street in which the segment is located. If an opportunity exists for a segment, the threshold value is lower than it would have been otherwise, and it is more likely that the model would recommend replacement. If the consequences are low and the cost of replacement is high, the threshold value tends to be high so that only a high probability of break for the segment will lead to replacement recommendation. On the other hand, the threshold value is lower for "high consequence" segments.

The segmental decision model described above can be applied to individual segments to decide whether or not a segment should be replaced at present time. We next address the determination of optimal replacement policies for the entire network under budgetary constraints.

The optimal policy for the network is to follow the results of the segmental decision model applied to each segment in the network. However, funds may not be available to follow the optimal policy (i.e., the budget may not allow replacing every segment that should be justifiably replaced). Under such a budgetary constraint, one tries to choose a portfolio of segments which would give the greatest benefit in the sense that replacing these segments would reduce the expected costs the most. Although selecting an optimal portfolio generally requires a complicated analysis, this problem has a particularly simple form which can be exploited to obtain an optimal solution which is also computationally simple. The solution ranks all segments needing replacement according to a priority ratio. Segments with a higher priority ratio should be replaced first and the cumulative cost is the required budget. Thus, with any given budget, the optimal group of segments that should be replaced is found. The priority ratio considers replacement costs as well as the risks for postponement, giving priority to high risk, low replacement-cost segments.

We will next illustrate the above approach by describing a case history. For the Gas Research Institute an integrated system of software programs, the Cast Iron Maintenance Optimization System (CIMOS), was developed to implement the segmental and network decision models. Figure 2 shows a flow chart of the major component of the CIMOS and input–output characteristics of each component. The three major components of the system are:

- Database of performance history
- Performance prediction models
- Optimization models

A brief description of each component is provided below.

Database of performance history

The performance history of each segment in the pipeline network is recorded in terms of the characteristics that influence the probability of (a) future occurrence of breaks and leaks, (b) migration of gas into a building given a break, and (c) a gas explosion given a break and gas migration. These characteristics can be divided into two categories:

Static characteristics that do not change with time (e.g., pipe size, length, and soil type); and

Dynamic characteristics that can change with time (e.g., cumulative number of breaks and leaks).

The specific static and dynamic characteristics to be included in the data base will depend on the types of information available to a company and past experience regarding the correlation of different characteristics with frequency of breaks and leaks. The performance prediction models and the optimization

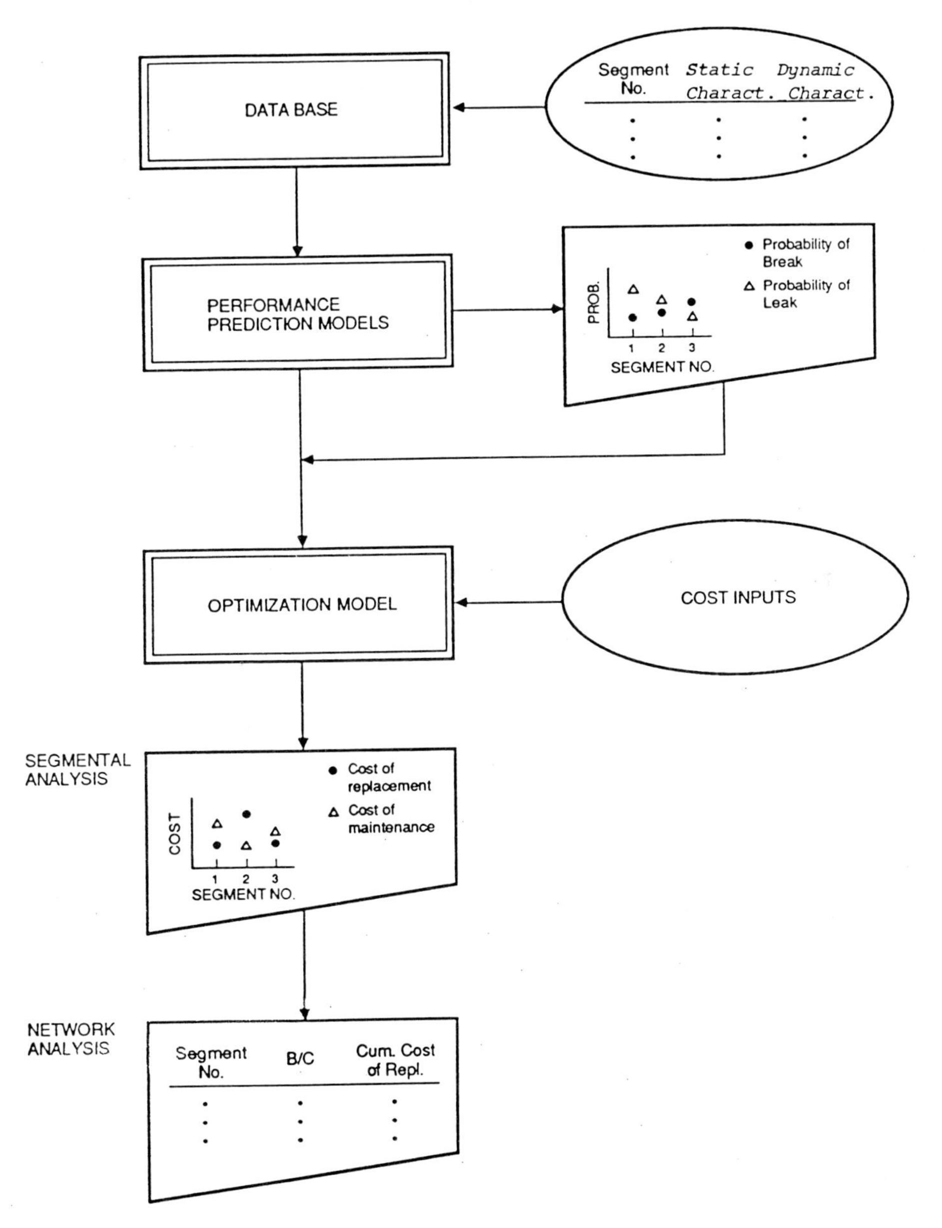

Fig. 2. A schematic flowchart of CIMOS.

models in the CIMOS are designed to be flexible so that they can be adapted for different types and amounts of data available to different companies.

In the implementation of the CIMOS for two gas utilities, the following static and dynamic characteristics have proven to be significant:

Static characteristics	Dynamic characteristics
Pipe diameter	Cumulative number of breaks
Segment length	Cumulative number of leaks
Operating pressure	Number of repairs in last 5 years
Soil type	
Traffic density	Age
Dept of cover	

Performance prediction models

Performance prediction models are developed to estimate probabilitites of breaks and leaks on a given segment during next year as a function of the static and dynamic characteristics applicable for the segment at the end of the previous year. A Bayesian diagnostic model that has been applied successfully in insurance and medical fields was used as the basis for developing CIMOS performance prediction models for this study.

The basic premise of this model is that the probability of failure (break or leak) on a specified segment can be obtained by adjusting the systemwide average probability of failure based on the relevant characteristics of the segment. If the proportion of failed segments with a certain set of characteristics X is greater than the proportion of all segments with characteristics X, then a segment with X will have higher than the systemwide average probability of failure. For example, suppose that 4 in. diameter pipes constitute 25% of the entire network, but 60% of the pipes that showed breaks were of 4 in. diameter. In this case, the probability of a break on a 4 in. diameter pipe would be higher than the systemwide average probability of a break. The objective of the model is to determine the characteristics which would significantly change (increase or decrease) the systemwide average probability of failure. By knowing the characteristics of each segment in the network, segment-specific probabilities of failure can be estimated and segments with a relatively high degree of hazard can be identified (see Kulkarni et al., 1986, for a mathematical description of the model).

Optimization models

Programs for the segmental and network decision models are included in this component of the CIMOS.

Inputs to the segmental decision model are the probabilities of breaks and leaks from the prediction models, and cost and policy inputs. The output consists of the total expected present worth costs of the two options as well as the major cost components. The decision whether or not the segment should be replaced now is identified with an indicator variable (1 indicating replacement, 0 indicating no replacement).

The input to the network decision model is the set of consequences and costs for individual segments obtained from the segmental decision model and the specification of available replacement budget. The output of the model includes a ranked list of segments in a decreasing order of the priority ratio and the cumulative replacement costs. For any designated budget, the decision maker can readily identify the segments that should be replaced, the priority order of replacement, and the associated benefits of such a policy.

Testing and validation

The CIMOS was tested for two gas utilities: Rochester Gas and Electric Company (RG&E) and Michigan Consolidated Gas (MichCon). Our approach was to use: (a) one data set to develop the performance prediction models and, (b) an independent data set to validate the model predictions and to evaluate the reasonableness of the CIMOS recommendations.

The size of the data bases used for the two gas utilities is shown in Table 1. The data used to develop the prediction models represented about a five percent sample of the entire cast iron piping system for each company. This is an adequate data base to estimate probabilitites even as small as 1 in 10,000.

Results of validating the performance prediction models and evaluating the CIMOS replacement recommendations are discussed in this section.

The performance prediction models estimate annual probabilitites of breaks and leaks as a function of the relevant static and dynamic characteristics of a pipe segment. In order to verify the reliability of these estimates, an independent set of random segments (which were not used in developing the models) was selected and the expected numbers of breaks and leaks on these segments were predicted for the four-year period of 1981 to 1984. The expected numbers of breaks and leaks were then compared with the observed numbers of breaks and leaks during this time period. The results of this comparison for RG&E are discussed below.

The independent data set used in testing the models for RG&E consisted of 207 segments. Table 2 shows the expected and observed number of breaks and leaks on the 207 segments for each of the four test years. A close agreement is

TABLE 1

Data based used in prediction models

Utility	Model development			Model validation		
	Number of segments	Performance history	Total number of data points	Number of segments	Validation period	Total number of data points
RG&E	700	1950–1980	21,000	207	1981–1984	828
MichCon	1850	1950–1985	64,750	204	1982–1985	816

found between the two sets. In addition to comparing the total number of breaks and leaks for the complete set of 207 test segments, it was necessary to evaluate the reliability of predictions for individual segments. If the models were working correctly, the segments predicted to have a high degree of hazard (i.e., a high probability of a break) should also show a high frequency of breaks and vice-a-versa. Statistical comparison indicated a close agreement between the distributions of expected and observed number of breaks over different intervals of the predicted probability of a break in both cases.

Evaluation of the CIMOS replacement recommendations

A total of 16 segments out of the 207 test segments was recommended by the CIMOS for replacement based on the economic evaluation made in the segmental decision model. The percentage of segments recommended for replacement:

- increases with pressure,
- increases with the number of past leaks,
- increases with the number of past breaks,
- increases when under the pavement (as opposed to when under the curb),
- decreases with size,
- increases with age, and
- increases with the number of pending leak orders.

These trends generally conform with engineering judgments. Consider, for example, past breaks on the 207 segments. The number of segments with past breaks of 0, 1, or more than 1, and the percentage of these segments recommended for replacement are as follows:

Past breaks	Number of segments (a)	Number of segments recommended for replacement (b)	% of segments recommended for replacement $(b/a) \times 100$
0	113	3	2.7
1	70	8	11
>1	24	5	21
	207	16	7.7

One can see from the above table that the percentage of segments recommended for replacement has increased with the number of past breaks. This is reasonable, since the probability of a future break would be expected to increase with the number of past breaks.

One example where the choice of a segment selected for replacement would not have been obvious is a 12 in. diameter pipe segment. Past experience indicates, and the break prediction model confirms, that the probability of a break on such a segment would be negligible. Based on the consideration of probability of a break alone, one would not have selected this segment for replacement. The reason the CIMOS selected this segment of replacement is economic; the segment had several past leaks and the expected number of leaks during the next year calculated from the leak prediction model is 1.4. The segment is under pavement and hence leak repairs would be costly. Also, there is some resurfacing work scheduled for this pavement. If pipe repairs are needed after the resurfacing work is completed, this would result in a penalty. Thus, the decision to replace the segment makes economic sense, since by replacing now, costly future repairs can be avoided.

Using subjective judgments and a manual inspection of available data (such as age, size, past breaks and leaks, and location), one may arrive, for some of the segments, at the same decision as those based on the results of the CIMOS. However, for other segments, subjective decisions may result in more costly actions than the decisions made on the basis of the economic evaluations of the CIMOS. The CIMOS provides a consistent basis for making optimal decisions for *all* the segments.

Potential cost savings from the CIMOS

In order to evaluate the potential cost savings that could be realized by using the CIMOS, the same set of 207 segments used in validating the performance prediction models was analyzed. Assuming that a specified replacement budget was available, different procedures may be used to decide which segments should be replaced. The impact of two alternative procedures on annual maintenance and repair costs was evaluated. The tow procedures were: priority ranking provided by the CIMOS and priority ranking by the probability of a break. RG&E in the past has used the ranking by probability of a break to choose segments for replacement. Since the actual maintenance and repair (M&R) costs for the 207 test segments were known, it was possible to find the costs that would have been avoided by replacing a particular segment. Figure 3 shows the reduction in the annual M&R costs (i.e., cost avoided by replacing the recommended segments) for different levels of the replacement budget for the two decision-making procedures. The use of the CIMOS shows a greater reduction in the annual M&R cost than the use of ranking by probability of a break. For a replacement budget of $100,000, the M&R cost for different replacement decisions would be as follows:

Decision	Annual M&R Cost ($)
No replacement	30,000
Replace segments using ranking by probability of a break	25,000
Replace segments using the CIMOS ranking	19,000

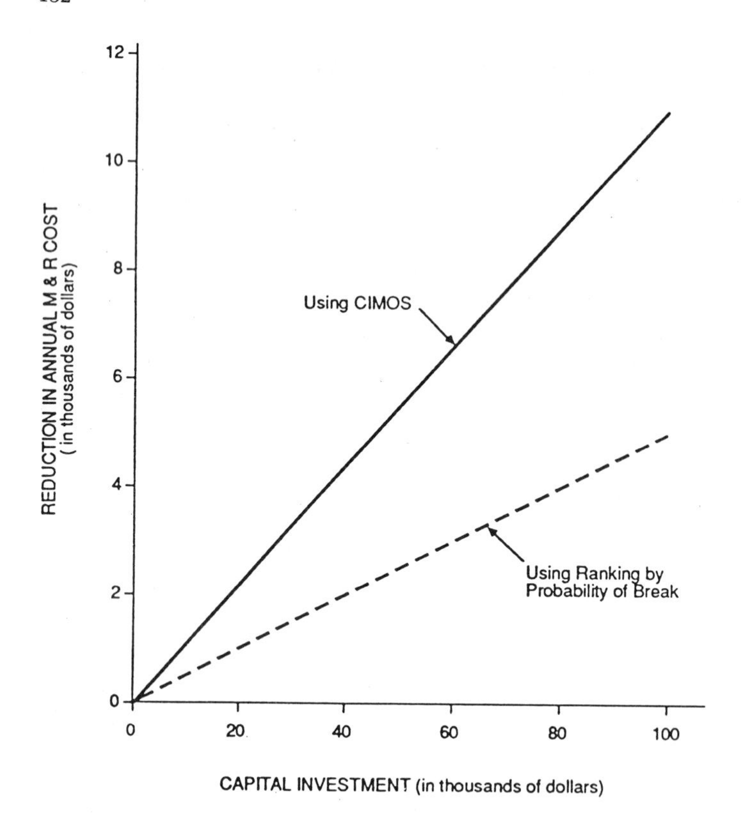

Fig. 3. Illustration of cost savings by using CIMOS.

Thus, the CIMOS would save ($25,000 − $19,000) = $6,000 in M&R costs compared to the other procedure; this is about [(6,000/25,000) × 100] = 24% savings over the other procedure.

Although the above analysis was only for a small portion (207 segments, or about 3%) of the total cast iron piping system in RG&E, similar savings of about 24% in annual M&R costs can be expected for the entire system.

5. SUMMARY AND CONCLUSION

The methodology outlined above provides a defensible and cost-effective replacement program for pipelines. The methodology can be used to:

- Estimate probabilities of breaks and leaks on different segments of pipeline systems.
- Evaluate which option is more economical: to replace a segment in the cur-

rent year or to continue maintaining it for another year with repairs as needed and then deciding whether or not to replace it.

- Find the optimal group of pipe segments which should be replaced, and their priority order, for any designated replacement budget.
- Select the most cost-effective repair method for those segments which would not be replaced.

The CIMOS provides an effective screening tool to rapidly evaluate all segments in a large network and separate them into two groups: (i) those which are candidates for replacement and hence should be investigated further, and (ii) those which are not justified for replacement at the present time based on economic and risk analyses and hence do not warrant detailed engineering investigation.

REFERENCES

Dugovic, G.R., 1980. There is gold in those records - sequencing the replacement of cast iron gas main sections. In: Proc. American Gas Association.

Gideon, D.N. and Smith, R.B., 1979. Analysis of the office of pipeline safety operations 1970–1978 reportable incident data for the national gas distribution companies. Report prepared by Battelle Columbus Laboratories for American Gas Association, A.G.A. Catalog No. X50281.

Kulkarni, R.B., Golabi, K. and Chuang, J., 1986. Analytical techniques for selection of repair or replace options for cast iron gas piping systems - Phase I. Topical Report, submitted to Gas Research Institute, Chicago, IL, September.

Kulkarni, R.B., Golabi, K., Dugovic, G. and Burnham, K.B. Optimal replacement decisions for cast iron gas piping systems.

Midwest Research Institute, 1982. Application of statistical techniques to gas operation. GRI Report No. GTI-81/0164.

Peters, J.W., 1983. Improve maintenance of cast iron mains. Pipe Line Ind., October: 39–42.

Short communication

Life cycle management of pharmaceuticals*

R. Bruppacher

CIBA-GEIGY, 4002 Basel, Switzerland

Life cycle management is a relatively recent concept in the pharmaceutical industry and is the consequence of losses due to product withdrawals following real or sometimes suspected safety issues. In the following I shall briefly outline the methodological aspects of risk assessment which have developed over the past decades to lower the risks both for the patients and for valuable products.

In all phases of development and marketing of pharmaceutical products a comparative assessment of quality, efficacy, safety, utility, need and cost effectiveness is mandatory. This assessment involves many disciplines: chemistry, biology, pharmacology, clinical medicine, epidemiology, economics and, increasingly, the social sciences. Political discussion is of growing importance and must be taken into account as well. A big issue has become the question of social acceptance especially in the field of psychotropic and endocrine pharmacotherapy as illustrated by the problems surrounding tranquillizers, stimulants, muscle building preparations or the "abortion pill".

The life phases of a pharmaceutical product can be divided in (1) chemical conception and synthesis, (2) biological screening (for efficacy), (3) toxicological testing (for safety) using animal experiments, (4) clinical trials (for efficacy/safety/modalities of administration) and (5) marketing. These phases have a complex and interdependent structure as illustrated by the plan of the development process in effect with some modifications for about 20 years at CIBA-GEIGY.

During all of these phases decisions have to be made. Uncertainties are unavoidable and sometimes lead to requests for more information instead of decisions. The need for a scientifically sound basis for decision-making has to be reconciled with the need for new medicines to treat and prevent disease. The situation calls for informed and intelligent interpretation of often controversial and contradictory evidence. Such a double-strategy can only be justified,

*Presented at the International Conference on Industrial Risk Management, Zürich, Switzerland, 16–17 January 1989.

however, if it is accompanied by surveillance systems which are able to detect the potential negative consequences of decisions very early.

Such systems of drug safety rely traditionally on four pillars. General pharmacological knowledge and experience, animal experiments, clinical trials and systematic collection and interpretation of spontaneous reports on adverse reactions of marketed products.

General pharmacological knowledge is important but no safeguard against surprises as shown by the examples of cimeldine, a powerful antidepressant, withdrawn from the market because of severe neurological side effects while the chemically closely related chlorpheniramine remains a widely used and well tolerated antihistaminic. Other examples include the non-steroidal antirheumatic drugs isoxicam and piroxicam or benaxoprofen and iboprofen.

The limitations of animal models for the prediction of (side-)effects in humans are also quite well known.

Clinical trials are as such a valid instrument for the assessment of the therapeutic potential and the relative safety of a preparation. However, this potential is achieved under conditions not likely to be reproduced during subsequent therapy in real life. Other drawbacks relate to small numbers that do not permit detection of rather rare adverse effects, to the relatively short duration of the therapy, so that cumulative effects are difficult to discover, and to the usually short observation period, which prevents the identification of effects with long latency.

Surveillance is both most important and most difficult in the marketing phase when products are used without the restrictions and close supervision that is customary during clinical trials. Spontaneous reports from physicians, pharmacists or patients can provide important signals, but do not lend themselves to quantification and risk estimation. This is easily understood if we consider that the number of reports depends not only on the incidence of the adverse reaction but much more so on the reporting rate, which is a function of the attitude of the physician, the climate of reporting in the respective country but also of the characteristics of the drug, of the adverse drug reaction and finally of the patient who is suffering from the adverse effect. Reporting rates, therefore, differ widely from country to country; in Europe the leading country with respect to reporting is Denmark with about 190 reports per 1000 medical practitioners per year, trailing is Italy with less than 2 reports per physician per year. Thus a difference of more than 100 fold can be noted.

Therefore, other surveillance schemes, e.g. intensive monitoring, post marketing surveillance studies and record linkage projects have been developed in the past two decades. All these schemes have advantages and deficiencies of their own. Intensive monitoring suffers from many of the same deficiencies as clinical trials. Post marketing surveillance and epidemiologic studies are open to numerous sources of bias and record linkage projects, linking billing data on prescription with data on hospital treatment, mainly lack information on co-factors and suffer from the usual problems of data collected for other purposes than for the ones aimed at in the epidemiologic analysis.

Despite all these difficulties all such approaches are necessary for an effective drug surveillance system, because only comprehensive and comparative information on usage and safety can save a product from allegations of unjustifiable therapeutic risk. Product information and marketing practices heavily influence the frequency and intensity of such allegations. Life cycle management must therefore comprise carefully coordinated pharmapolitical, commercial and scientific decision-making.

Integrated safety*

Christian Jochum

Safety Department, Hoechst Aktiengesellschaft, Frankfurt am Main, F.R.G.

ABSTRACT

Jochum, C., 1990. Integrated safety. *Journal of Occupational Accidents*, 13: 139–144.

Safety considerations are an important part of chemical company's corporate culture. It is now widely understood that safety should be integrated in the total life cycle of chemical and biochemical products and processes.

Hazards of these products and processes are based on material properties. Therefore they can be identified by scientific methods right at the beginning of product life. The most efficient (and generally most economic!) approach is to choose safe chemicals (or microorganisms) and inherently safe reaction conditions. As far as this cannot be achieved, the risk has to be controlled by safety systems (construction, instrumentation, organization).

During the early phase of product life (laboratory, pilot plant) usually the process design changes (e.g., variations in auxilliary materials, reaction conditions). During the whole product life cycle the knowledge of hazards will improve. Therefore systems have to be implemented to monitor these changes and to assess their impact on risk management. These systems must take into account also the "human factors" during design, operation, use and disposal of products and processes.

The principles of integrated safety are successfully used since many years by the Hoechst AG. Important structures to implement this system are guidelines, organizational means and training. The responsible managers in R&D and operations as well as in sales and waste disposal get advice and assistance by safety specialists coming from various disciplines.

INTRODUCTION

Learning from mistakes is and always has been an important aspect of safety work. Much in our accident prevention regulations is the result of analyzing an accident. This must not be seen in a negative way. Most of the natural sciences were originally based on observations. However, predicting developments is a greater stimulus to man than describing facts. In sciences, this striv-

---

*Presented at the International Conference on Industrial Risk Management, Zürich, Switzerland, 16–17 January 1989.

ing led to revealing conformities with natural laws. Those working in the field of industrial safety, on the other hand, had to be content with the description and subsequent correction of mistakes for a very long time. The anticipation of hazards before they arose was left to the intuition of individuals. It is quite astonishing that in such a vitally important field as hazard prevention, the methodical looking ahead, the purposive search for relations between cause and effect began at such a late stage. There is little point in asking whether this was a result of the lack of importance, attached to safety consideration in earlier years, or whether this deficiency was the reason for the low standing formerly accorded to safety experts.

Since some decades we experience a growing awareness for the risks of industrial installations. Dedicated scientists and engineers started to look into safety of processes as well as into quality, yield or reliability.

These individuals were not satisfied with simply describing a risk that had become obvious, or with preventing it from resulting in damage. As important as the safety valve of a chemical reactor certainly is, they want to investigate instances when it might never be actuated. This, however, assumes that one is familiar with the laws underlying the risks. Therefore, they followed a path which logically and consistently has led to the concept of integrated safety. The inherent logic of this concept is so stringent that a number of major chemical companies have developed independently similar systems.

Our safety concept is based on five theses:

1. Risks inherent in chemical processes are based on material properties and can be determined by scientific methods at the initial stages of process development.
2. The identification of hazards and the development of safety measures are an integral part of every planning process.
3. Risk assessment has to cover the product life from research over production and use through to disposal.
4. Controls are essential and must be ensured by means of organizational measures.
5. The human being in his function as planner, operator and checker is an essential part of every safety concept.

These theses are doubtless undisputed. However, it is vital that the conclusions drawn from them are also put into practice. They cannot and must not be allowed to peter out in statements of a purely programmatic nature. If one has to implement these theses, the line supervisor responsible for safety is entitled to get clearcut rules and effective aids. It is against this right that our concept of integrated safety, which I should now like to describe to you, must be measured. It combines and continuously updates elements which have proved themselves in our company for many years.

THESIS 1

The first step in any risk evaluation is the identification of hazard sources. These are, if we concentrate on chemical hazards, strictly related to the substances. Right at the start of the process development it is, therefore, possible and also necessary for the undesirable properties of the materials and the planned chemical reaction to be determined in addition to the desirable ones. This not only serves to protect our staff in the laboratories. The earlier the hazards are identified, the more effectively, and often also more easily, protective measures can be integrated into the planning process. Particularly at the start of a process development it is often quite possible to switch to safer materials or reaction conditions.

It belongs to the good features of the numerous safety regulations in Germany that they enforce this principle. According to our Ordinance on Hazardous Substances ("Gefahrstoff Verordnung") we not only have to evaluate the hazards of the substances but also to check if the use of less dangerous substances is feasible.

This principle can be applied to biotechnology, too. There it does not mean using safe chemicals, but harmless microorganisms. Up to the last few years the type of microorganism was determined by the desired product, e.g., a vaccine or an antibiotic. Genetic engineering now enables us to deliberately choose microorganisms which are intrinsically safe for both human beings and the environment. This is not the place to discuss the safety of genetic engineering in detail, but this advantage certainly deserves more attention.

In chemical companies it is a good practice since many years to emphasize the intrinsic safety of using safe substances – both for safety and economic reasons. Our young scientists at the universities, however, are still scarcely taught this principle. I really wonder what must happen to make the academic world more aware of these vital facts!

In order to identify hazards, it is above all necessary to have first-rate information. We have established our own data processing system which covers ecological, toxicological and other safety related properties of substances. We use outside data bases, too. For most of the data that may still be lacking we have our own testing facilities.

For each project the data should be compiled in a material data report. This document should accompany each process development right from the beginning and be continuously updated.

In the course of the process development, the requirements in respect of data completeness increase continuously. The basic data required at the start should be clearly identified in the material data report. They are ultimately supplemented to the data volume needed for the final licensing applications.

THESIS 2

Identification of sources of hazard is, however, only a part of the risk assessment. The risk is defined as a function of the extent of damage and of the probability of it occurring. The possible extent of damage is determined by the material properties and the technical conditions, e.g., the pressure resistance of apparatus. The probability of occurrence, on the other hand, is very largely determined by the technique used. Here, for example, I think of the reliability of control units or the possibility that sparks or other ignition sources may be present. This consideration shows that the actual risk assessment cannot be detached from the process engineering side of the process development. The risk involved in each process engineering option under consideration is assessed. If it is unacceptable, the process engineering must be changed and the risk assessed anew.

The engineering controls available for chemical processes are so advanced, that even extremely hazardous substances can be handled safely. However, it certainly makes more sense to strive for intrinsic safety rather than using sophisticated controls. Again I believe that responsible use of genetic engineering will enable us to use "mild" biochemical processes instead of the sometimes "hard" classical chemical procedures in quite a few cases.

In the iterative process of assessing risks and determining protective measures, even experienced planners must be able to consult qualified safety experts. Here it is not a narrow specialization that is called for, but as wide a view as possible, both of the field of chemistry and that of process engineering. It is scarcely possible for these requirements to be combined in individual persons. Our solution is a close cooperation between chemists, engineers and physicists in the Safety Department. Furthermore, there is quite a lot of specialized knowledge in other departments of our company which can contribute to safety. The safety experts have to be aware of this and use this across all fences between departments, which sometimes are pretty high even in the same company!

THESIS 3

The risk assessment of chemical substances must not stop at the producers fence. According to Germany's Chemicals Law since 1982 every substance which we start to sell has to be checked in detail especially with regard to it's safe application. Many other countries, including all members of the European Community, have similar regulations. Problems still arise from the many substances which had been on the market before these regulations became effective. Many initiatives are now in place to fill the gaps of knowledge. However, limited resources in toxicology and ecology are slowing down this process. Therefore, it is imperative to set priorities.

THESIS 4

No doubt, the awareness for safety and ecology becomes very high in chemical industry. However, we need control mechanisms. Their details strongly depend on the company's organisation and the local regulations. It is essential that the integration of safety considerations in the process development is guaranteed. We use a set of controls:

The project team itself has to check the safety at characteristic points such as the transition from laboratory to pilot plant and further to large scale production. We therefore provide a specific protocol and expertise input, if necessary. A key feature is that the person assuming responsibility for running the process at a larger scale decides whether the safety data are sufficient. If he does not feel comfortable, it is not be him who must determine the necessary data, but the submitting department. This is doubtless the right way, both from the legal and psychological aspects.

At the advanced planning stage an even more detailed risk assessment is necessary. Now the inclusion of the Safety Department is mandatory. Finally we have to submit a detailed application for licensing. The respective German regulations have been aggravated since many years. As an aftermath of the Sandoz accident we have now to submit for the majority of our chemical plants — even for the existing ones — a formal safety report which comes close to the demands for nuclear installations. Above all, public hearings are part of the licensing procedure. This alltogether means that in Germany we now need 2–3 years to get a license for a chemical process, whether it is new or only modified. We consider this as a serious drawback in comparison with other countries.

The risk assessment, resulting in the safety concept, is just as dynamic as the process itself. It must be adapted to all the process engineering changes which occur during the lifetime of a process, as well as to new safety and toxicological findings. Regular safety inspections of the plants can only guarantee this in certain points. Of more decisive importance is the continuous trustful contact between operations and specialized departments, in particular the Safety Department. This cannot be regulated, but must be actively developed by the specialized departments. They thus require qualified staff who are competent contacts for the operations superiors, from the foreman to the plant manager and plant engineer.

THESIS 5

An integrated, i.e. comprehensive, safety concept would be incomplete without the involvement of people. Our plants are not operated by robots. The people working in our plants not only have a right to a safe job, but are themselves an essential element of our safety concept. The training of our staff, the provision of information on possible hazards and safety measures, clear instructions, and the motivation to perform their tasks in an attentive and safe manner are important elements of any safety activity. The gratifying drop in the accident figures especially in recent years can largely be attributed to this branch of safety work.

CONCLUSION

Safety is an essential element of corporate culture. More than ever before the acceptance of the chemical industry depends on the trust placed in it by our employees and neighbours. Low accident figures alone do not suffice. We must show that we are continuing to utilize every opportunity to further increase the safety of our processes and workplaces. Safety consciousness must be fully integrated in the day-to-day activities. We no longer have any need to convince anyone of that. Rather it is necessary to further improve the methods used. The integrated safety concept, of which I have highlighted some of the principles, pursues this objective. It covers the whole span of a process from the laboratory over large-scale production through to use and disposal, from material data through to the use of psychological methods in the motivation of staff.

Safe operation of chemical plants
Methodology and practice of risk prevention*

R. Gasteiger
Bayer AG, Leverkusen, F.R.G.

ABSTRACT

Gasteiger, R., 1990. Safe operation of chemical plants. Methodology and practice of risk prevention. *Journal of Occupational Accidents*, 13: 145–156.

Safe operation of chemical plants is not only a question of operation procedures. Safe operation possibility starts very early during the design phase of a chemical plant where all the risk factors can be evaluated, and adequate techniques against those risks can be found out.

Principally, there are only a few typical risk factors like fire, explosion, chemical reactions and toxic risks for environment or personnel. Those risk factors can be looked for by systematic methods and – if necessary by some repeating steps – effective measures against those risks can be evaluated.

It needs a very consequent planning which is strongly correlated to the steps of risk analysis and even a strong quality assurance in course of the phase of erection of the chemical plant.

Nevertheless, a strict management during the phases of operation and, especially, the phases of maintenance, repair or changes in design is necessary.

An overview of the instruments of safety analysis is given, and the way by which risks are eliminated, is shown by an example.

1. PROBLEM

In recent years a number of major disasters have shaken the chemical industries, e.g. Flixborough in 1974, Sandoz in 1986, which characterize the time interval and the magnitude of the accidents. There had been also less dramatic incidents during the same time period and unfortunately, these continue as well. But we are very hopeful to minimize both in future – the frequency and the magnitude of incidents. I think this is not only a hope but a reality at least for the chemical plants in the Federal Republic of Germany. This can be shown by the tremendous efforts on the fields of safety techniques and the high stan-

*Presented at the International Conference on Industrial Risk Management, Zürich, Switzerland, 16–17 January 1989.

dard of safety technology reached in comparison to other European and non-European countries.

The chemical industry needs reactive materials for preparing their products. Often those materials can only be handled under certain process conditions which means that only a very small variation in temperature, pressure, composition of components, etc., is allowed to prevent for example a runaway-reaction. On the other hand often the products or components are toxic or polluting and by this dangerous for personnel or environment.

Safe operation of chemical plants is not only a question of operational procedures. Safe operation starts very early during the design phase of a chemical plant where all the risk factors can be evaluated and adequate techniques against those risks can be developed.

An analysis of the accidents having occurred in the past shows that in all cases there would have been enough knowledge and information available to the persons responsible to avoid those incidents. Even in the most severe cases – for example as the Bhopal accident in 1984 – it had not been a lack of knowledge of safety technology but it had been wrong decisions for operation in connection with a disregard of safety units.

This means that during the step of detailed engineering the results of a parallel ongoing safety analysis should be strongly used and even during the phase of erection of the chemical plant a strong quality assurance is necessary to bring differences between planning and hard-ware realisation to zero.

Nevertheless, a strict management during the phase of operation including the phases of maintenance, repair and especially modifications in design is necessary to assure that the standard reached in safety technology at the time of start-up of the plant is not violated. In contrary this standard should be continuously increased by adopting the experiences of operation to the safety concept of the plant.

2. TYPICAL SOURCES OF RISK GENERATION

Principally there are only a few typical risk factors like fire, explosion, chemical reactions and toxic risks for environment or personnel after leakage of products.

These final incidents are the back-end of a sequence of less or more events starting with an initial failure. That is why safety analysis is primarily looking for failure possibilities. A list of those failure possibilities is given in Table 1. This list is not intended to be complete, but it is illustrative.

Those failure possibilities can be classified in groups which are correlated to physical, mechanical, chemical phenomena like

- increase of pressure
- build-up of explosibility conditions
- wrong concentrations

TABLE 1

Failure possibilities

Pressure increase	- wrong dosage rate - closed valves - blocking of pipes
Explosibility conditions	- loss of inert atmosphere - remaining product after cleaning - dry operation of pumps
Changes in concentration	- wrong dosage rate - non-uniform mixture - mistake in product sequence
Leakage of product	- malfunction of relief valve - untight valves or flanges - corrosion
Temperature increase	- malfunction in cooling system - exothermal reaction - closed valves on pressure side of a pump
Interaction of materials	- reverse flow of products - wrong sequence of products - wrong quality or composition of products

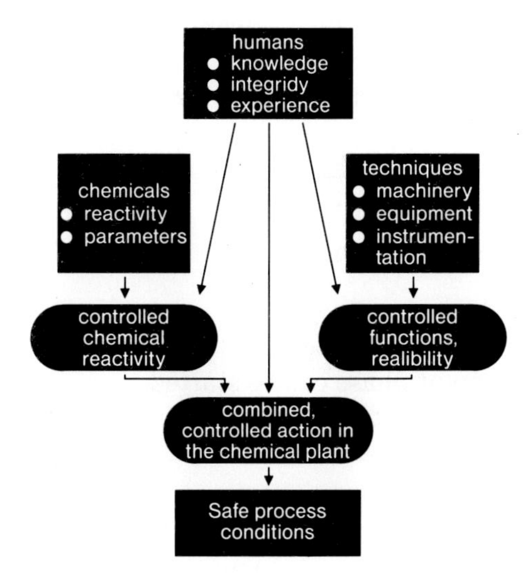

Fig. 1. Basic concept for safe chemical plants.

- leakage of products
- increase in temperature
- interaction of materials

which may lead to one of the above mentioned final incidents.

It is a broad field of work to find out the possibilities for initial failures and the identification of events which follow as a sequence and which may at last lead to an incident. This work can only be done by a systematic study of all the dependencies between physical/chemical behavior of products, technical equipment, process operation and last not least human influences. This means that for safe chemical plants the three factors
- personnel
- chemistry
- techniques

must interact together in the right way (Fig. 1). Not to forget, there are a lot of requirements to the acting personnel, mainly with respect to knowledge, experience and personal integrity.

3. METHODOLOGY FOR RISK IDENTIFICATION

In praxi there is the problem of bringing all existing knowledge and experiences on potential sources of hazards into a systematic order by a formalised methodology. This methodology should accompany the steps of realisation of a chemical plant from the beginning of laboratory tests until the start-up of production. For realisation this needs a team of specialists which covers all the disciplines involved.

The interaction by a team of specialists has the high advantage that the different views and experiences of persons from for example safety departments, planning departments, and production departments are brought closely together.

Typically for each step of development the same steps of work must be done in a cyclic way (Fig. 2)
- systematic analysis
- valuation of risk potentials
- design of technical measures against risks

Normally these steps are repeated several times until adequate technical measures are found out. Then the next development phase can be started.

We have strongly formalised the way of proceeding with the safety related actions (Fig. 3). This figure shows that there are basically four main steps
- sampling of basis safety data
- conceptual safety design
- systematic safety analysis
- approval of safety installations

On the left hand-side of Fig. 3 one can see the steps of development for a technical plant. On the time axis one can see the correlation between technical steps and safety relevant work. After each of the four steps there is a stop where the results must be certified by the involved working groups.

Now, what is the contents of the four steps in detail? Basically each step

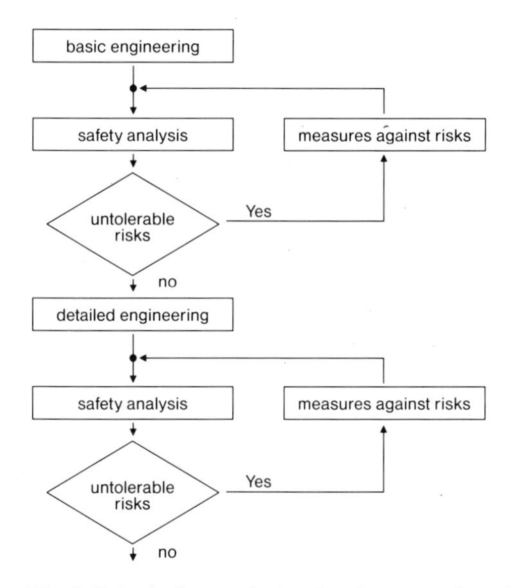

Fig. 2. Principal steps during development of a safe chemical plant.

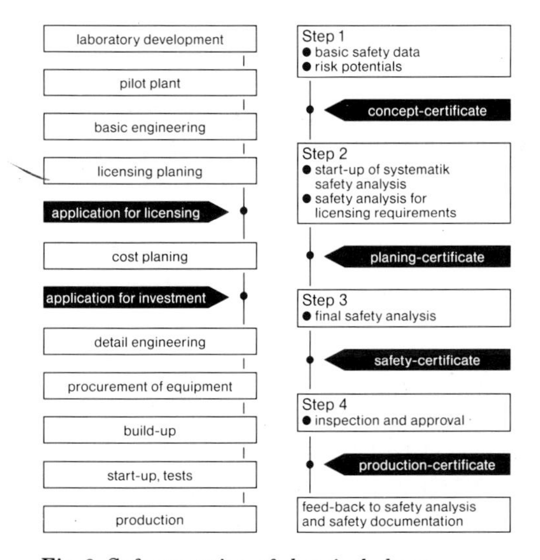

Fig. 3. Safe operation of chemical plants.

follows the same pattern: possible risk potentials must be evaluated and technical measures against those factors must be formulated.

Sampling of basic safety data

During process development on laboratory scale a systematic sampling of data for identification of potential hazards from material and reaction behavior is done. If necessary those data must be measured in laboratory to complete the basis of knowledge.

Looking for safety data means systematic search for properties which can induce hazards. This means that besides the necessary knowledge of the toxic properties of materials or mixtures investigations on physical/chemical reaction behavior have to be carried out.

Concerning the prevention of development or propagation of fire this means the determination of

- flammability of materials
- fire sustaining properties of materials
- ignition of materials
- burning properties and fire damage of materials
- ...

For the field of explosibility conditions this means the determination of

- lower/upper explosion points, explosion limits
- detonation properties
- inertization media
- maximum explosion pressure
- ...

A broad field of necessary investigations are the thermal aspects of process design. These investigations are obligatory to find out possible runaway-reactions. This for example includes the experimental determination of

- thermal stability of materials/mixtures
- thermal and mechanical stability
- heat generation under normal reaction conditions
- heat generation and decomposition possibilities in case of deviations from normal reaction conditions
- ...

Experimental methods like DTA (Differential Thermo Analysis), ARC (Accelerating Rate Calorimetry) or Power Compensated Calorimetry can be used (Figs. 4 and 5) to find out the thermal behavior of pure materials or mixtures. These investigations are carried out under regular and even under simulated accidental process conditions.

This knowledge is completed and assured during process development in the stage of a pilot plant. Especially the interaction between structural materials and products, the dimensioning of the process equipment under safety considerations and the formulation of the instrumentation requirements – in some cases supported by testing – is completed.

This step is finished with the "*concept-certificate*" which means that from point of safety view all necessary data are available to assure a safe design.

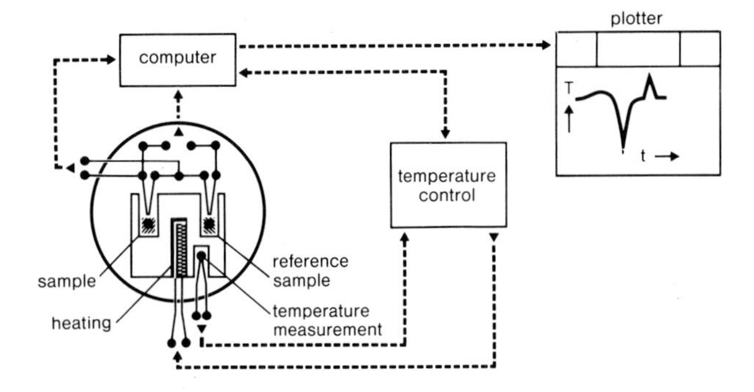

Fig. 4. Differential Thermo Analysis (DTA) principal diagram.

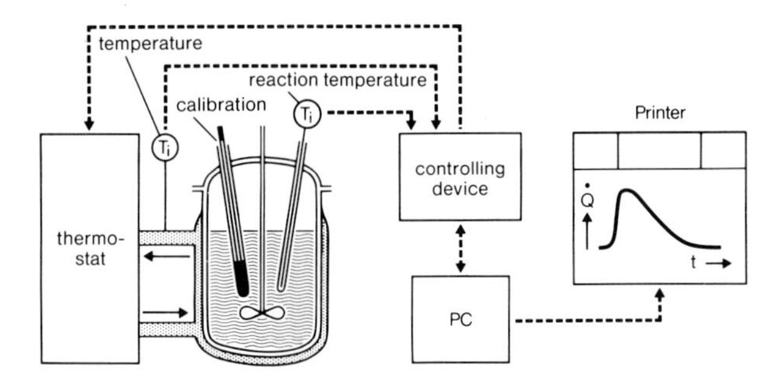

Fig. 5. Computer-controlled calorimetry principal diagram.

Conceptual safety design

Parallel to the phase of basic engineering a safety analysis is started. This work is first done in a very rough manner by using the so-called "hazard area method".

We have developed for the application of this method special lists with questions for the different hazard areas

- physical risks
- risks by energy release of fire, explosion or exothermal reaction
- health effects
- environmental effects

Those lists are not to be used as checklists but their application helps to be systematic. Nevertheless, it is necessary to analyze the specific process conditions.

As an example the following gives an impression of the kind of questions of those checklists for case of energy release.

Checklist Energy Release

1. Are the materials thermal stable under process conditions

2. Do the following conditions lead to undesired reactions or exothermic decompositions?
   change in process conditions
   deviation of the material composition
   access of impurities
   catalytic reactions
3. ...

In reality those lists are very detailed and they represent our experiences in practical plant analysis.

As a result of this first stage of safety analysis the basic safety requirements and the possible solutions can be fixed. This means that the way for detail engineering is cleared, the blue prints, descriptions and analyses for licensing can be completed and the cost planning can be detailed for the necessities of application for investment.

This step is finished with the "*planning-certificate*" by the members of the involved working groups. This means that all necessary data and informations are available to start with detail engineering.

Systematic safety analysis

Parallel to the phase of detail engineering the systematic safety analysis is completed. This work is done with an ongoing interaction to the work of the planning staff.

During this stage the already mentioned hazard area method or, if necessary, a more sensitive operational analysis method is used. All these methods work also in a cyclic manner, where first for hazard potentials is asked, then an evaluation of the possible consequences follows. This cycle ends in the proposal and design of technical measures against those hazards and looking again for satisfying action of the equipment (Fig. 6). This is done for all the process segments following the P&I diagram. At first it is necessary to briefly explain the desired function of the equipment, machines, instruments and the corresponding pipelines within the process segment being checked. Then the systematic search for the possible deviations of the desired functions can be started. If postulated causes cannot be excluded, the theoretical disturbance or the postulated failure must be assumed to be possible.

For the judgement, whether safety equipment is necessary or not, the possible consequences of the postulated failure are evaluated without considering existing technical measures against this failure. This is necessary to get a feeling of the magnitude of the respective consequences. On that basis the type of safety system and the necessary reliability can be fixed.

By repeating the explained steps including the technical corrective measures it can be checked whether the action of the safety equipment is satisfying. All

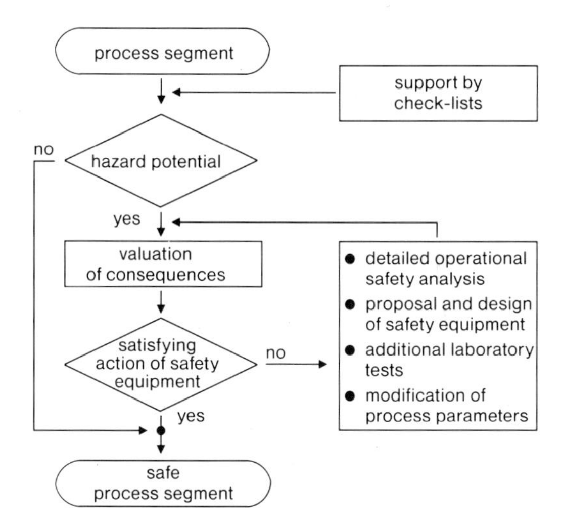

Fig. 6. Principal decisions during safety analysis.

this safety analysis work is to be finished before procurement of the equipment to assure satisfying function in terms of safety.

The written results of each step of the safety analysis in total gives the documentation of the safety concept of the chemical plant. These papers are also useful documents during the forthcoming phase of production. They should be used as a basis for decisions if modifications in process or plant design are to be done and they are also a helpful instrument for phases of maintenance or repair.

This step is finished with the "*safety-certificate*" by the members of the involved working groups which means that with respect to safety requirements the designed plant can be safely operated.

Approval of safety installations

During the phase of erection of the plant a lot of components are to be inspected due to the existing regulations.

Additionally to this – let me say "official action" – a check is made to approve existence and well-function of all the safety equipment which was decided being necessary as a result of the safety analysis steps.

This step is finished with the "*production-certificate*" which means that from point of safety view the testing phase and start-up of production can begin.

4. AN EXAMPLE FOR RISK ELIMINATION

Imagine we use for the production of a polymer-product a reaction vessel of some cubic meter volume. This vessel is used batchwise. The reaction needs two components, A and B. An activator is necessary to start up and keep on the polimerisation.

A batch production is started with filling in about 10% of component A which is then heated up to reaction temperature of about 120 °C. In the following controlled flows of components A and B and the activator are added. Stirring of this mixture is necessary to avoid accumulation and to keep the solution homogenous. The generation of reaction heat is compensated by cooling the reaction vessel with water.

For safety considerations the following facts are of importance:

- the used materials are toxic, which means that a release of these substances to environment is to be avoided
- the heat production during the step of polimerisation is in a range that there should not be an accumulation of product A and B, because delayed dosage of activator will cause heating-up accompanied with increase of pressure. Only a small amount of non-polimerisated part of component B (about 10%) is allowed to stay within the range of pressure resistance of the vessel.

In reality this process is a little more complex, but let me terminate this for demonstration purposes at this stage.

The first information we get from this rough safety analysis on the stage of a "hazard area analysis" is that technical safety measures are necessary because of the toxicity of the products which on release will lead to intolerable consequences for personnel and/or environment.

As you will immediately realize there is more than one possibility of coping this problem. A relief valve which opens to the environment is not allowed because of the toxicity of the products. But there would be the possibility of a blow-down system to a stand-by tank. A solution like this means high investment costs and – what is much more important – there must be a satisfying off-gas system and additionally a solution to the waste problem if this system would ever be used.

Another possibility – and I am sure you can easily find more – is to install a safe flow control for the correct dosage. If there is an accumulation of component B in case of a delayed dosage of activator, there may result a risk of too high energy release. The sequence of failures can be compensated by a differential control for both flows.

As you can see in Fig. 7 this flow control is installed for production purposes in connection with the computer-controlled instrumentation system. Because in our example the correct action of this system is safety relevant, it is for us a principle of design to have more than one system for safety action, and one of those systems must not act in connection with the computer-controlled instrumentation system but it is to be installed in hard-wired design.

Figure 8 gives you the result of our safety analysis (for this part of the problem). The control for flow B and for the activator is doubled. Each of the

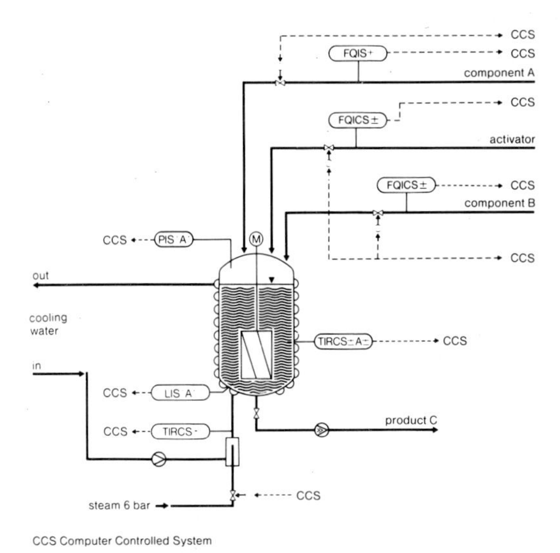

Fig. 7. Reaction system before safety analysis.

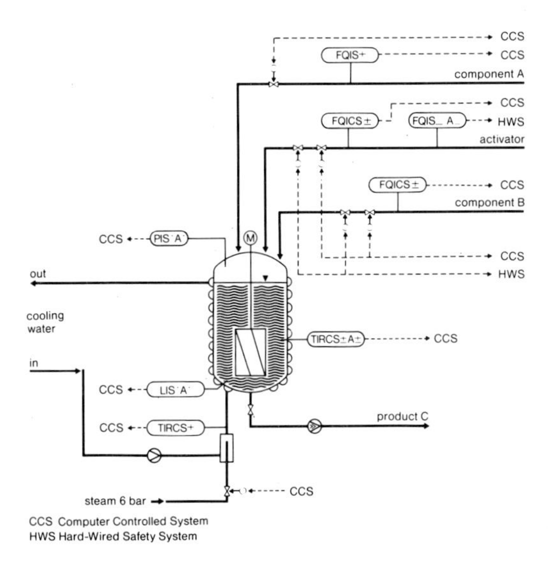

Fig. 8. Reaction system after safety analysis.

controller acts on separate valves. One controller is supported by the computer-controlled instrumentation system the other is installed in hard-wired design.

All these discussions should be completely documented to have the full information for the production phase. But this information is also necessary for phases of maintenance, inspection or repair.

You see that a step-wise analysis of chemical processes can lead to technical solutions which are optimized and satisfying both in terms of production parameters and safety consideration.

5. CONCLUSIONS

Safe operation of chemical plants starts very early during the design phase of the plant. As we know from experience there is enough knowledge and information available to avoid incidents.

The safety-relevant questions can be cleared by a simultaneously ongoing safety analysis which is accompanying the different steps of planning. This safety analysis must be systematic and well organized. In praxi it is useful to support the preparation of a safety analysis by a checklist for special hazard areas like fire, energy release or environmental effects.

Good engineering practice combined with cross-checks by safety analysis assure safe plant design.

In addition efforts should be made to keep differences between plans and reality near zero. This goal can be reached by a strong quality assurance during the phase of erection of the plant.

Nevertheless, a strict management during the phase of operation including the phases of maintenance, repair and especially for modifications of process or plant design is necessary. Even for this phase the results of the safety analysis are helpful because they give the complete information about the safety concept of the plant.

We are on the way to realize these actions very systematically for our plants and I am sure this is the best way to minimize both in future – magnitude and frequency of incidents.

The safety of the Rapid Transit System in Zürich*

Peter Zuber

Bauabteilung der Generaldirektion SBB (Swiss Federal Railways), Bern, Switzerland

ABSTRACT

Zuber, P., 1990. The safety of the Rapid Transit System in Zürich. *Journal of Occupational Accidents*, 13: 157–165.

The Rapid Transit System in Zürich is a regular, modern train which will be constructed with the most modern safety equipment. Nevertheless, the thought that 100,000 persons a day will gather at one given place, all underground, under houses and rivers was uncomfortable. A team of experts, combining worldwide knowledge on hazard analysis methodologies, specific knowledge from the Swiss Federal Railways and a great deal of imagination produced a detailed catalog of hazards. Train accidents and fires were identified as the major risks. The layout of the whole construction and a lot of specific safety measures were submitted to a careful evaluation, resulting in both an acceptable protection level and an optimal cost/benefit ratio.

1. THE NEW RAIL CONSTRUCTION UNDER THE CITY OF ZURICH

The Rapid Transit System in Zürich will become operational at the end of May 1990. After this date, fast trains will circulate on practically all of the train tracks in the city of Zürich and its surroundings at least every half hour. A new highlight in Zürich will be the new double decker trains. This has been made possible thanks to a new 12 km underground track which will avoid a bottleneck of the railroad at major intersections and will almost double the capacity. Heavy traffic, namely the whole regional traffic of the Swiss Federal Railway will circulate here. Trains with up to 1200 passengers will circulate every $2\frac{1}{2}$ minutes during rush hour. The underground extension of the Zürich train station, the so-called Museumsstrasse train station (Fig. 1), will accommodate the largest number of people. A four track through station with two platforms will be located underground (there is no space on the ground level). The

*Presented at the International Conference on Industrial Risk Management, Zürich, Switzerland, 16–17 January 1989.

Fig. 1. Cross section of the Museumsstrasse train station.

station is located under the Museumsstrasse between the Landesmuseum and the central station (Hauptbahnhof). Its altitude is determined by the Sihl and the Schanzengraben: the two watercourses have to cross the platform floor of the train station and therefore lie on the same level as the pedestrian floor which is linked by escalators, elevators and stairs to the platform of the currently existing train station.

The train station Museumsstrasse is designed to accomodate 7000 passengers every 10 minutes or 100,000 passengers per day. Approximately the same number of people circulate at the Stadelhofen train station which is covered by a heavy arcade.

The train station Stettbach which is located on the out-skirts of the city will accomodate very few arriving and departing passengers but nevertheless, it is still for the most part located underground. Tunnel stretches of 1.3–5 km long are located between the stations and two longer viaducts are located outside Stettbach.

2. THE TASK OF THE SAFETY EXPERTS

The Rapid Transit System in Zürich is a regular, modern train which will be built according to high performance requirements. This means that it will be constructed with the most modern safety equipment such as train surveillance and radio communication. Therefore there is no reason to worry about the safety of the Rapid Transit System.

Nevertheless, the project leaders asked themselves at an early stage whether the new rail construction could generate hidden hazards: the thought that such a large conglomerate of people gathered at one given place, all underground, under houses and rivers was uncomfortable.

Two worries remained:

1. The worry of building something wrong, such as a trap where passengers cannot find their way out or the thought of a small fire breaking out which could all of a sudden have catastrophic consequences. Mistakes in the construction of underground structures can rarely be made up for later. International newspaper reports unfortunately confirm that such worries are impossible to ignore.

2. A further worry that is difficult to neglect is that someone could require exorbitant safety measures such as escape shafts or tunnels running parallel to the train tunnel and ask the unfair question "will you take on the responsibility if something goes wrong?"

In order to avoid this problems, a team made up of international and external experts (known as a Safety Council) was formed which was given the following tasks:

1. to role play in their mind all the imaginable accident scenarios and to examine with what measures they can be avoided or, if this cannot be done with safety measures, examine how the consequences can be controlled better;

2. to suggest an optimal "packet" of safety measures, that is, one that has a maximum effect with the minimum expense. Safety means the protection of the human body and human life.

3. DRY ANALYSIS, RICH IDEAS FOR SAFETY MEASURES

The cooperation of worldwide knowledge on hazard analysis methodologies, specific knowledge from the Swiss Federal Railway (SBB) and a great deal of imagination and creativity resulted in a detailed catalog of hazards listing possible threats the SBB could be involved in.

The probability of an accident occurring in the Museumsstrasse train station is higher than for example an accident occurring in a normal suburban train station with 10 times fewer trains and 100 times fewer passengers per day.

The accident analysis showed that train accidents and fires are the major risks for the Rapid Transit System. Figure 2 summarizes the distribution of the risks in the Hirschengraben tunnel.

Before the safety measures were defined, the Safety Council assumed that the construction and technical structures were correctly engineered and exe-

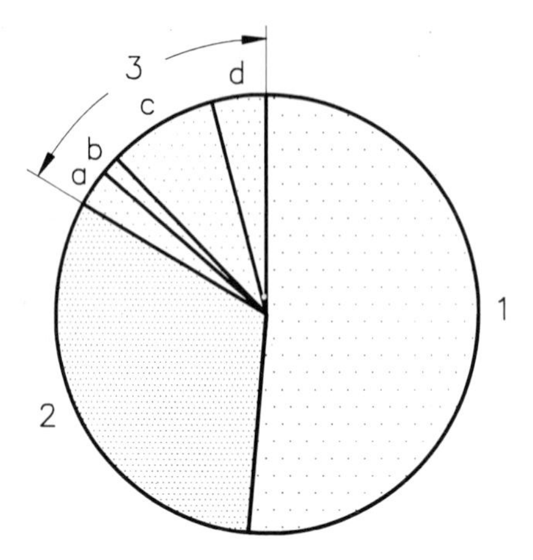

Fig. 2. Distribution of risks in the Hirschengraben tunnel: 1 train accidents, 2 fire, 3 others (a: sabotage, b: crime, c: personal accidents, d: work accidents).

cuted. The council made sure that the classical engineering problems were being taken care of by other departments and concentrated on safety problems that were not taken into consideration yet.

4. CHOSING THE RIGHT SAFETY MEASURES OR HOW SAFE IS SAFE ENOUGH

All the measures to improve the safety of the Rapid Transit System which did not involve high costs were implemented. The decision on implementing safety measures which were costly was more difficult to make. It was taken on the following basis:

1. according to existing regulations

The decision is easy to make if certain safety regulations already exist. For the safety questions that had to be dealt with here, there were very few defined codes or regulations. In this case, international regulations – especially from Germany – were consulted. This was of little use. Various safety regulations which exist in other countries cannot be applied to the Rapid Transit System in Zürich. For example, in Germany the regulations require that in every underground railway system, there should be an emergency exit at every 800 m. This may be worthwhile on flat terrain but in Zürich emergency exits would have to be built in areas with an altitude difference of 180 m. This does not serve any purpose. Rather, it is very expensive.

2. according to good engineering judgement

If no regulations exist, the decision on which safety measures to incorporate is based on the state of the technology, the opinion of the experts and the systematic analysis and assessment of risks. We will see that this can be done in various ways.

3. according to the comparison with existing systems

- an overall safety level comparable to existing railways should be warranted.
- the safety level of the various track stretches should be as equal as possible.
- the safety measures in the areas of construction, equipment and organization as well as engines and coaches should be at a similar level.
- the safety measures chosen should have an acceptable cost/benefit ratio.

The result of this analysis is the following:

The two highly experienced consulting companies chosen, both safety experts, used different hazard analysis methods yet both delivered similar risk reduction recommendations.

The recommendations delivered by one company were driven by the protection level. The other company used cost/benefit observations.

Recommendations

Recommendations driven by the protection level

The likelihood of any hazard was estimated and classified under six levels, ranging from frequently to impossible. Likewise, the team classified the effect of any hazard in four categories ranging from negligible to catastrophic. A graphic illustration of the risk profile can be seen in Fig. 3. Every hazard is represented by a numbered point. The further point lies in the top right hand corner of the graph, the more serious is the risk. If the point lies in the bottom

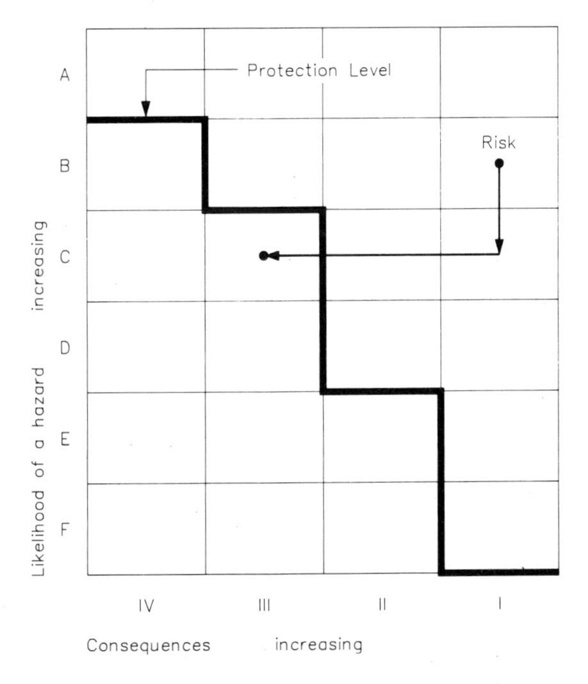

Fig. 3. Risk profile.

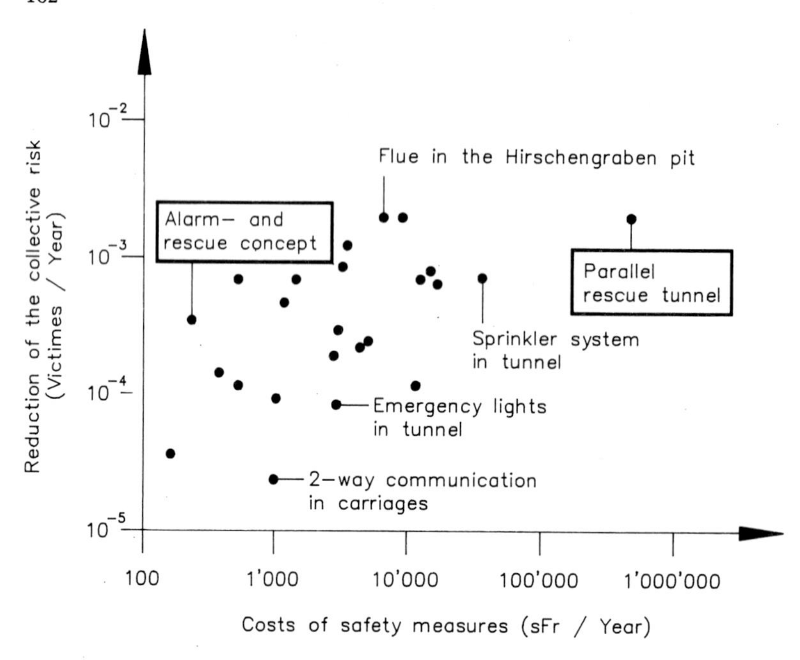

Fig. 4. Cost of safety and risk reduction measures in the Hirschengraben tunnel.

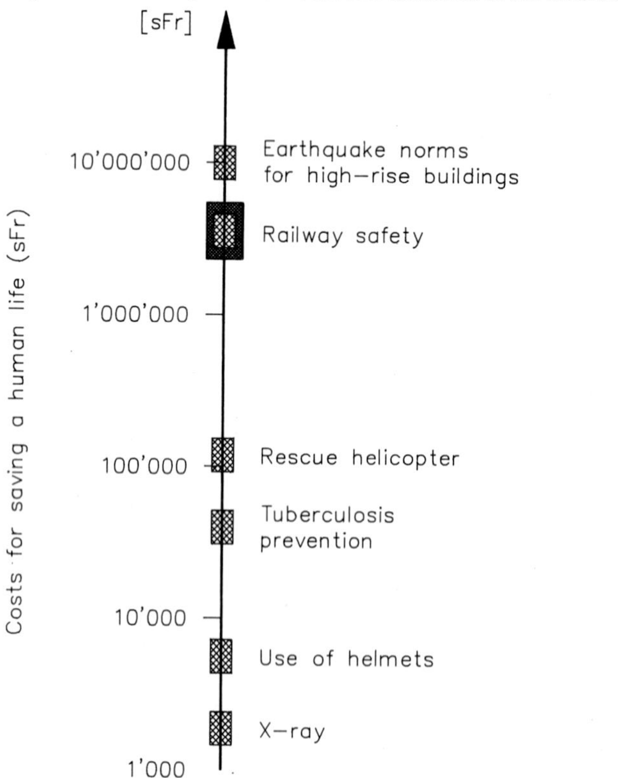

Fig. 5. Costs for saving a human life.

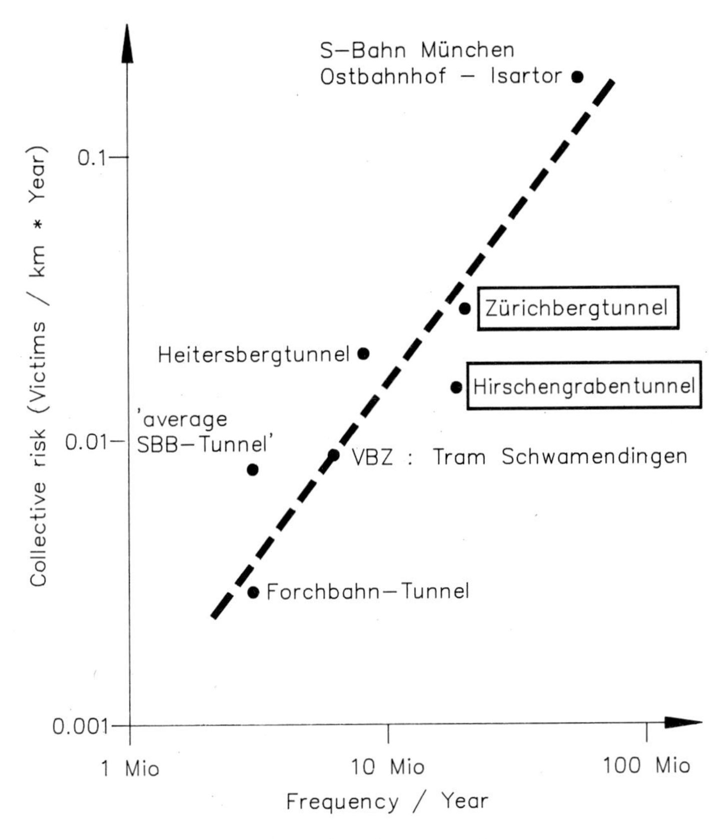

Fig. 6. Collective risk of different railway tunnels.

left hand corner of the graph, the lower the risk. Somewhere in between lies a limit where a risk can be classified as acceptable. Consequently the protection provided can be considered sufficient. This limit is called the protection level.

A protection level based on personal judgement is not calculable. It is placed where the experts believe that the safety of human life is sufficient and acceptable. The experts in this case were members of the safety council. Apart from their field expertise, they were also concerned with the safety expectations of the public since they are the ones which operate the Rapid Transit System.

Risks that lie above the desired protection level do not meet the necessary safety standards. Therefore they require additional measures to reduce the risk. If several risk reduction options were generated the most cost worthy would obviously be chosen.

This method is easily understood and can be further expanded. A newly identified hazard and the resulting necessary safety measures can always be

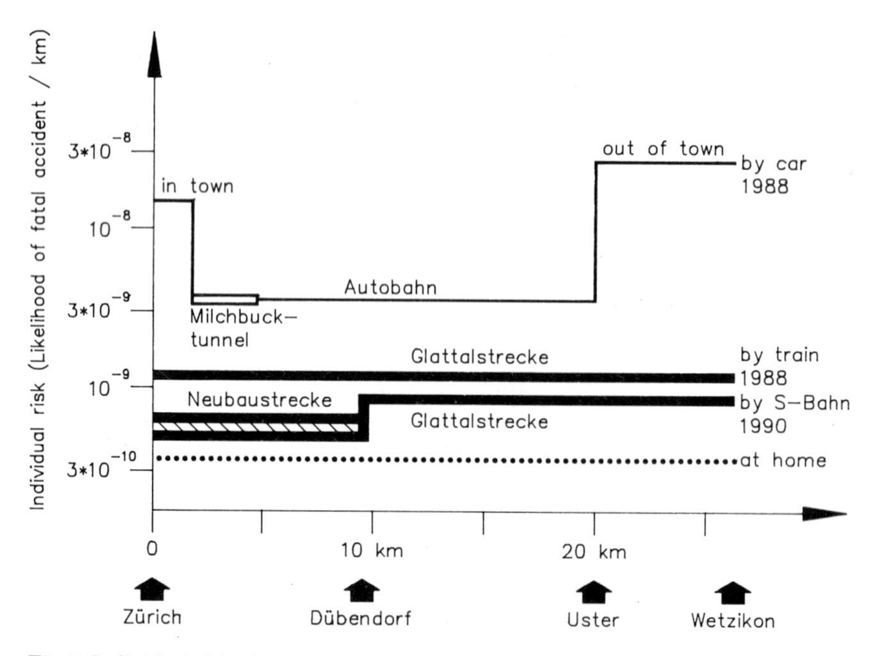

Fig. 7. Individual risk of a railway passenger compared to others.

added to the lists, as has been the case so far and will continue to be until the Rapid Transit System becomes operational.

Since this method is based on relative assessment, it allows for relative instead of absolute comparisons.

Recommendations using cost/benefit observations

The *cost* of implementing the safety measures was calculated based on yearly costs, whereby the cost for investment and maintenance were also taken into consideration.

Benefits achieved by the implementation of these safety measures were defined as human lives that could be saved. In this context, the potential of severe accidents resulting in fatalities was more heavily weighted.

Figure 4 indicates the cost of safety and risk reduction measures in the Hirschengraben tunnel.

The measures that were investigated show large differences in cost and benefit. Certain measures show large risk reductions for low cost, but others bring little improvement despite high costs.

For example, with the safety measure known as the "emergency preparedness" concept,approximately 600,000 Sfr.have to be spent, to save one human life. The other option is to construct a separate parallel rescue tunnel for more than 100 million Sfr. to save one human live.

With this method, quantified criteria for choosing an optimum safety package can be achieved. The figures are only relevant when they are linked together with sufficient statistical data. The method allows for comparison with other railway stretches of this Rapid Transit System and with other existing railway. As expected, in view of the relatively high safety level of railway systems, the costs for saving a human life are relatively high – adding up to millions of Sfr.

Figure 5 illustrates the measures used to save a human life vary according to the area of application and to the safety measures implemented. If the results of the safety measures of the Rapid Transit System were to be expressed statistically, we can see that it is worth all the effort. The overall risk is below average. A person traveling alone takes advantages of the additional safety efforts made to avoid catastrophes and travels safer with the Rapid Transit System than with a conventional train.

Figure 6 shows the comparison of the overall risk at the Hirschengraben and the Zürichberg tunnel with other tunnels. The magnitude of the risks is in most cases determined by the size of the number of people traveling at a given time.

Figure 7 indicates that the involuntary risk of a person riding in a train from Zürich to Wetzikon will decrease when the Rapid Transit System becomes operational in 1990.

In order to make a comparison, involuntary risks in a car and in a household are shown.

CONCLUSION

The deep interest in risks makes us more aware of the critical parts of a complex structure such as a railway system. One of the direct consequences of the reflection on safety lies in the implementation of a special surveillance and maintenance plan for safety critical elements of our structures.

Quality: how to meet customer demands in a process from design to disposal*

Lars Johnson
Volvo Car Corporation, Göteborg, Sweden

ABSTRACT

Johnson, L., 1990. Quality: How to meet customer demands in a process from design to disposal. *Journal of Occupational Accidents*, 13: 167–170.

High quality will be of paramount importance for the future in the automotive industry. Quality is a customer-oriented concept and the whole organisation should have a customer-oriented attitude. This implicates the need for efficient information systems to feed back customers' preferences and demands. In the future knowledge about customers' experience of the products through its entire life cycle will be even more important.

1. QUALITY – A COMPLEX CONCEPT

High quality will become the prerequisite for survival for products in the automotive industry. By high quality, we mean that all the intrinsic properties of the product, which create the concept of quality, must be of high class and must meet our customers' requirements. We refer primarily to properties such as reliability, safety, comfort and "value for money". There are other properties as well, such as performance, ergonomy, road holding, appearance, detail workmanship and ease of service which are also important. Customer tolerance for properties which do not live up to the expected level will be very small in the future.

One of the reasons for this is that car production is going to be greater than sales for several years to come, and competition is going to get tougher. In addition, the basic properties of the cars on offer will be quite similar because of increased legislation on environmental and safety aspects, etc.

When the customer chooses the "right car" in our market sector, a low product price at the moment of sale is not going to have a decisive influence on his choice of car. On the contrary, the overall view is going to become more dom-

*Presented at the International Conference on Industrial Risk Management, Zürich, Switzerland, 16–17 January 1989.

inant. For example, product reliability must be of the highest class. Otherwise, it is the customer who suffers considerably as result. Apart from the loss of confidence caused by the breakdown, it causes loss of working time and other inconvenience.

2. IN FUTURE – A DEVELOPED QUALITY CONCEPT

It is obvious that the concept of quality is going to become more and more sophisticated in the car industry. Apart from the actual quality of the car, such properties as the "software" will continue to be given greater importance. These additions to the product, to give higher quality of function, will put higher demands on the marketing organisation and on the personal competence of the people who support the use of the product, thus giving the customers genuine security of function.

In order to further reduce the time that the customer needs to spend on routine maintenance, it will be increasingly necessary in future to reduce the need for regular maintenance. In its turn, this will further be the requirement for high quality in all the car's systems.

Such system quality cannot, however, be built up without improving the quality of the components in the systems to the same degree, as well as the quality of the sub-components in the systems. However, these systems, subsystems and individual components include both material properties and human labour, which is a tangible and possibly decisive factor in creating high quality.

In other words, individual human competence, which is built in to the actual product over and above the levels which must be included in the peripheral service, then also becomes a survival factor for a product as complex as a car.

This complexity also means that we must work systematically, at all stages of the development and production process, to achieve high quality. So let me spend the rest of my time telling you about how we work to achieve high quality, both in all the details and also in the sum total of all the products and service, seen through the customer's eyes.

3. QUALITY – A CUSTOMER-ORIENTED CONCEPT NEEDS EFFICIENT INFORMATION SYSTEMS

It is necessary to see everything with the customer's eyes, and not with the narrow gaze of the specialist, in order to achieve overall quality and avoid the far too narrow trade-offs which are usual in western quality work. This is often based on the prevailing belief in a universal "correct" quality, without any feeling for the dynamic development, which goes on in an environment which seems ever more demanding to us.

A constant high rate of development is therefore necessary and the level of quality, which was optimum at a specific instant, will quite certainly be totally insufficient in a few years or even a few months.

In order to find out what high quality is in the customer's eyes, we have developed two measurement instruments, which are called VOTE and VOICE.

The VOTE survey is performed by inviting a number of presumptive customers to drive a number of competitor's cars in accordance with a pre-arranged schedule together with a passenger who is trained in interview techniques. Before, during and after the drive, the passenger asks the "test driver" a number of previously formulated questions, which elucidate the driver's evaluation of various aspects of the car being driven. The answers, which are collected from several markets, are then evaluated. In this way, we can formulate relative scales for the properties of the car in question, compared to the Volvo car. This helps us, in its turn, to allocate priorities in product development.

For customers who have already bought a Volvo car, we make similar surveys after longer or shorter periods of Volvo ownership. We call this VOICE – the voice of the customer.

This is done in our largest markets – U.S.A., England and Sweden and also Germany, Italy and Japan. In this way, we find out about the properties which have been received positively, neutrally or in a manner which is not satisfactory for the customer. We have to give the latter highest priority in product development in order to get a satisfied customer who will buy a Volvo again, next time he decides to buy a car. In this way, we also get indications about what the customer thinks about our software in conjunction with service, plus any after-market additions to the car after the actual purchase.

To this knowledge of what the customer wants, you naturally have to add the properties which have perhaps passed un-noticed by the car-buying public, but which indicate future needs or requirements from the authorities in various markets. The aforementioned properties of the car can be said to add up the *primary quality of our products.*

4. THE SECOND STAGE

If these choice of properties represent the first stage then naturally, "perfect" workmanship down to the smallest detail in each car, is an equally important stage in getting high quality in the product. How do we achieve this second stage?

Briefly, this is done by subjecting the systems, sub-systems and individual components to exhaustive development and testing activity in order to ensure perfect function in all service environments and conditions of loading. In addition, comprehensive development work is done to make sure that the systems can be produced, installed and the function checked in the cars, so that *all* cars *always* meet the specified requirements with a margin. We call this the zero-defect philosophy. This is increasingly necessary to preserve and reinforce our competitiveness.

The foundations of the zero-defect philosophy are the attitudes of managers, foremen and operators.

It must mean a genuine will to always personally guarantee high and even better quality in all work.

In addition to this, advanced technology must be developed to ensure high quality of manufacture and assembly, despite such factors as difficult working environment, monotonous work operations and continuously increasing demands for speed and precision. These demands, together with demands for high quality of assembly must in many cases be met by comprehensive mechanization and/or robotization, since human ability is no longer sufficient.

In low-stress areas of production, it can however be possible to achieve high quality by giving people advanced equipment of various sorts, to provide a better working position, or make work easier in other ways.

In order to achieve zero-defect in these work operations, there must be constant striving to devise product designs which make incorrect assembly impossible, for example. This is what is popularly called "fool-proof design".

For the actual manufacturing of components for the cars, new methods must be developed, so that thousands of "zero-defect" components can be produced. One such method, which is increasingly used to reduce the variation in the dimensions and properties of manufactured components is *statistic control* of suitable processes. There are processes which are not only controlled to product components within specified tolerances, they are also controlled so that, to the greatest possible extent, they produce "ideal" components, and aim at meeting an ideal value – *a target value.*

In its turn, this requires that old tolerance theories should be abandoned. In return, considerably higher function quality, general product quality, customer satisfaction and cost efficiency can be attained, thus increasing competitiveness.

5. CHANGES - LEADERSHIP AND ATTITUDES

Built to these efforts is an ever increasing requirement to get to grips with a broad spectrum of interconnected problems of labour, technology and, not least of all, personal motivation at all levels. In other words, it is a question of changing basic attitudes. Our new attitude must be that every complaint or fault in the product is unacceptable and must be eliminated. Examples of new orientation in quality can be seen by their thousands in industry after industry, and will be increasingly prominent in our industry.

What I have just said gives a few short glimpses of the quality development activity which is decisive for the survival of our products in the "car war". Choosing a customer-oriented quality concept entails an immediate requirement for effective information systems. These must quickly and "intelligently" relay the customers' real and undistorted attitudes to our cars and service. The information must be transformed to decisions for action in structured and wise courses of action. Despite all systems, we must not forget that the decisive factor is the common striving of many individuals to achieve their utmost to ensure the survival of our products in a saturated market.

Digital switching system EWSD: operation and maintenance using expert system techniques*

Max Sevcik
Siemens-Albis AG, Zürich, Switzerland

1. INTRODUCTION TO OPERATION AND MAINTENANCE FUNCTIONS IN DIGITAL SWITCHING SYSTEM EWSD

The Siemens EWSD is a digital switching system for a world-wide use in many different areas starting from a small decentralized rural switch up to a very large central office handling vast number of trunks and subscribers.

Flexible configuration of *system functions* according to the needs of a particular administration (e.g., PTT) is another important principle. EWSD functions can be varied from standard telephony switching service to sophisticated Integrated Services Digital Network (ISDN), capable of handling speech and digital data switching functions simultaneously (Ribbeck and Skaperda, 1984).

Apart from actual switching functions, the EWSD performs many supporting tasks helping the administration's personnel to run the system. These tasks

*Presented at the International Conference on Industrial Risk Management, Zürich, Switzerland, 16–17 January 1990.

are called *operation and maintenance functions* and may be informally defined as follows:

(a) *Operation functions*

All functions supporting the installation of new system components like trunks, subscriber circuits (hardware *and* software), etc. Similarily all functions used to reconfigurate and to modify existing system components belong to this category as well.

(b) *Maintenance functions*

Maintenance functions assure correct performance and high availability of the whole system. They are responsible for automatic supervision of all principal system components and for recovery actions needed to circumvent problems. As a result, the maintenance engineer will receive the necessary advisory information about the fault and the possible corrective actions.

Figure 1 illustrates the principal EWSD system components. The complexity of the EWSD system and the diversity of the operation and maintenance functions demand for a precise and flexible *command language* (Fig. 1) in order to facilitate the necessary man–machine interaction by PTT's administration personnel. EWSD complies fully with the respective CCITT recommendations (CCITT, 1988). The CCITT Man Machine Language (MML) defines generic syntactical and semantical rules, governing command languages for switching systems. Its *particular* use for *specific* operation and maintenance functions in EWSD environment has to be documented in a set of *system-specific* operation and maintenance manuals which are to be delivered and configured with each

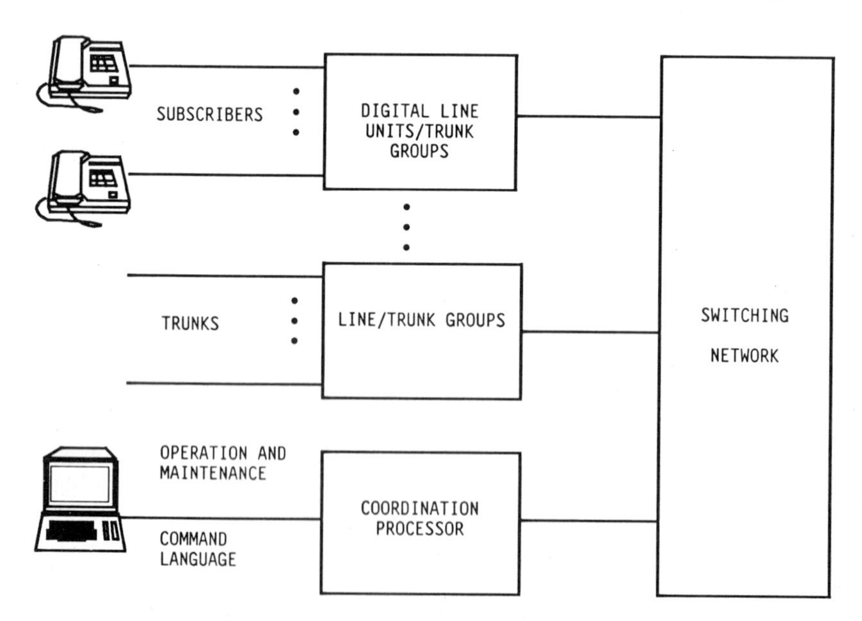

Fig. 1. EWSD digital switching system components.

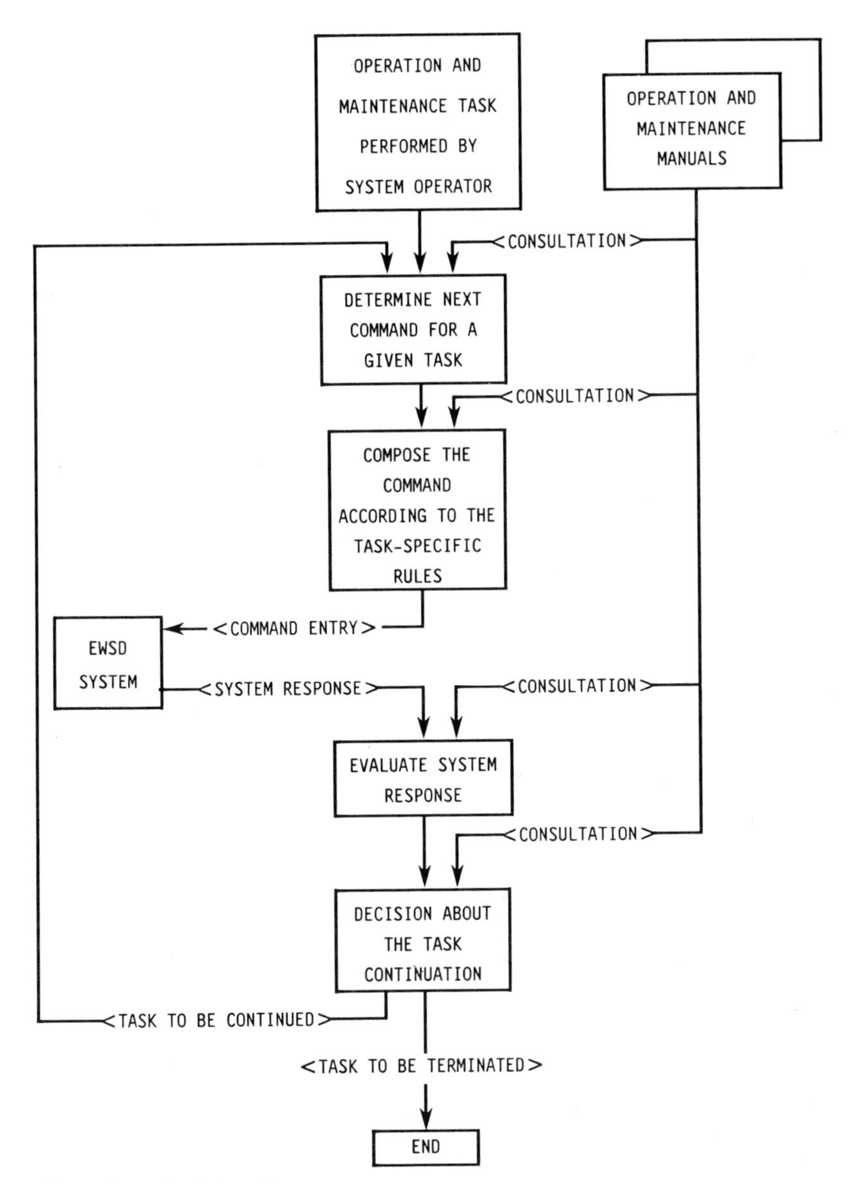

Fig. 2. Operator interaction.

exchange. Currently this type of system documentation comprises of several thousands pages (depending on system configuration).

Even though the EWSD automates many elementary operation and maintenance functions, the system operator has to consult frequently the operation and maintenance manuals, specifically when he/she is asked to perform more complex *tasks* consisting of many elementary man machine commands with associated system response. Figure 2 illustrates schematically the interaction between a system operator, manuals and the EWSD switching system during an operation and maintenance task (in CCITT terminology: operational procedure).

2. POTENTIAL RISKS INCURRED BY OPERATOR INTERACTION

Operation and maintenance functions influence the overall state of the switching system. In this respect, similar operational requirements known from other complex technical entities like power plants, power distribution, traffic supervision apply to digital switching systems as well.

There are, however, some important differences in the assessment of risks due to errors incurred by human operators:

(a) *Very high system availability*

The availability is measured as total cumulative downtime during the life span of the system. For switching system, extremely high availability, e.g., one hour total downtime in 25 years or more is required. This implies, that all extentions, changes and reconfigurations have to be performed on *live system*. Manual actions by human operator (Fig. 2) may in case of erroneous decisions, misinterpretation of manuals, etc., *immediately affect* the traffic in parts of or in the whole communication network.

(b) *Impact of system malfunctions*

Switching system malfunction in a communication network (e.g., due to human operator failing to comply with the standard operating tasks rules) does not lead to physical damages in the system, buildings or even injuries as it may happen with "conventional" technical system. System errors or downtime cause first of all losses of revenue for the PTT administration (customers find their lines "dead", "busy"; therefore will not complete their calls and cannot be charged for it) and for some customers as well (e.g., electronic fund transfer, point of sale systems). Such financial losses may even exceed the installed value of the switching system itself, should the impairement of service last for longer time. The quality of service as observed by customers is important in order to maintain the communication networks trustworthy in the long run, thus creating the potential for future traffic expansion.

3. IMPROVED CONCEPT FOR OPERATION AND MAINTENANCE TASK INTERACTION

Major improvement (e.g., reduction of human operator errors) in the current approach may be achieved if the stepwise execution of an operation and maintenance task would be automated, i.e. in minimizing or even eliminating the discrete interaction steps. As suggested by Fig. 2, the subdivision of a task in separate steps is basically due to the *consultation activity* in the operation and maintenance manuals. To automate the task means primarily to dispose of the manual consultation steps.

Siemens-Albis has adopted the approach of replacing the consultation steps by techniques of formal knowledge representation of the operation and maintenance manuals content. This kind of representation is suitable for an automated interpretation in a computing system commonly called *Operations System*, thus virtually eliminating the need for a human operator to consult the manuals.

Figure 3 presents the key functional components in the new concept as compared to the conventional approach. The requirements for this novel system may be more precisely stated as follows:

(a) The discrete interaction steps for all operation and maintenance tasks

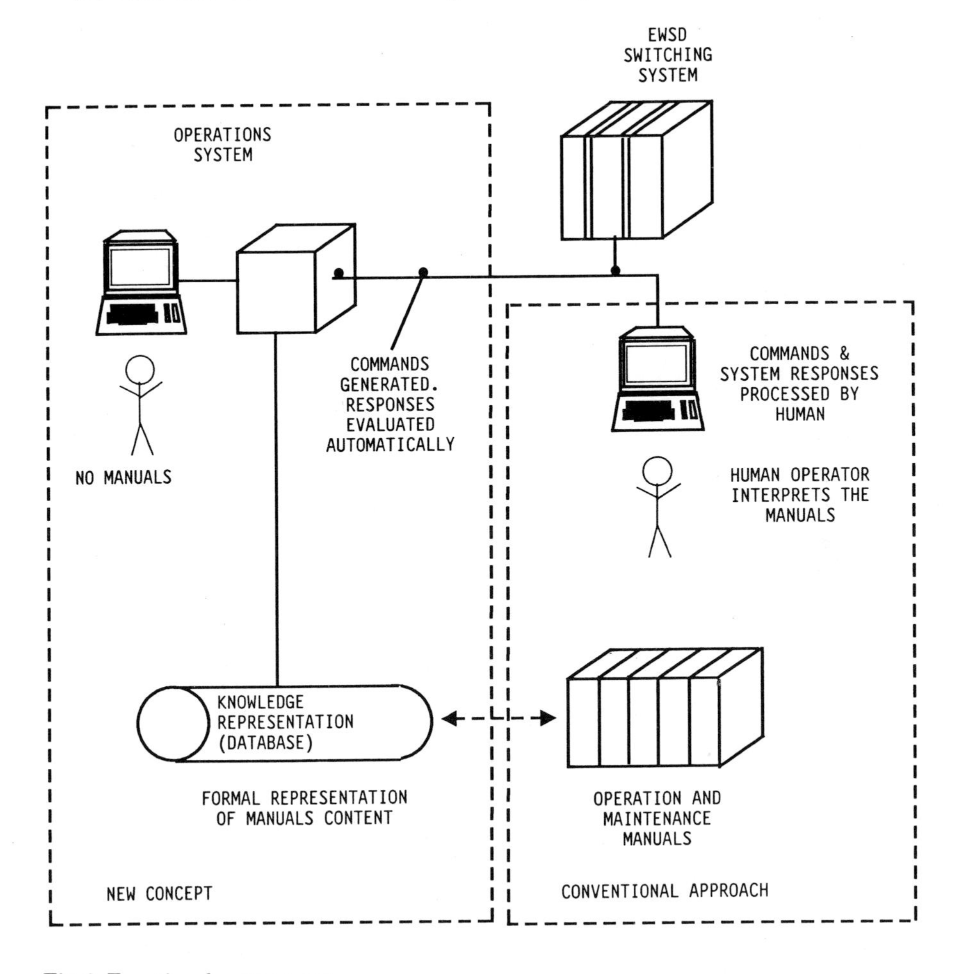

Fig. 3. Functional components.

will be selected automatically by the operations system according to the formal rules *derived* from the manuals and stored in the knowledge database.

(b) All rules describing the *individual commands* and system *responses* have to be maintained and interpreted by the operations system (e.g., facts about the command syntax, value ranges of parameters).

(c) Information requested from the human operator must be kept at minimum (being error-prone) and be limited to situations where formal rules are not feasible (e.g., manual actions like physical replacement of a faulty circuit board).

(d) If requested by the operator, the operations system must supply full explanation for every interaction step, command and system response in accordance with the system state (e.g., dependent on the state of EWSD system components and previous interaction steps within the same operation and maintenance task).

(e) Easy adaptation of the formal rules in the knowledge database must be provided for in order to reflect ongoing changes and extensions in the EWSD systems attached to the operations system. Special precautions are necessary to maintain *data consistency* between the operations system and the EWSD systems.

(f) For emergency purposes the conventional "paper" operation and maintenance manuals have to be produced and maintained consistent with the operations system as well.

It is obvious that the formal knowledge representation is one of the crucial steps in this undertaking. Several concepts from expert system and knowledge engineering domains have been successfully exploited in this approach.

4. IMPLEMENTATION

Basic components in the new operation and maintenance concept of Fig. 3 imply the need to implement two functions:

(a) *Online computer system (expert operations system).* Its principal task is to interpret the formally represented content of the manuals in the knowledge database and to *interact online* with the EWSD switching system using regular command language (CCITT, 1988) of the same type as in the conventional approach where the human operator handles the command language on his own.

(b) *Offline tools to maintain the knowledge representation database.* The content of conventional operation and maintenance manuals has to be transformed into a *formal language* (interaction rule system, its principles are beyond the scope of this short contribution) by an expert operation and maintenance engineer. Software packages (editors, database products) are necessary in order to support the initial transformation and the future updates in the knowledge representation database.

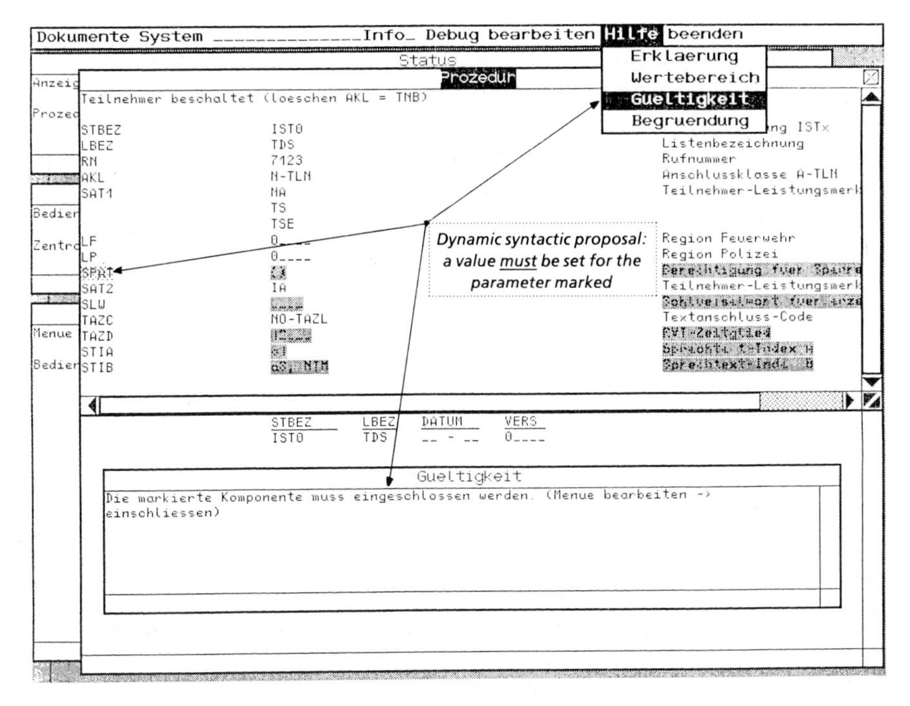

Fig. 4. Sample screen layout 1.

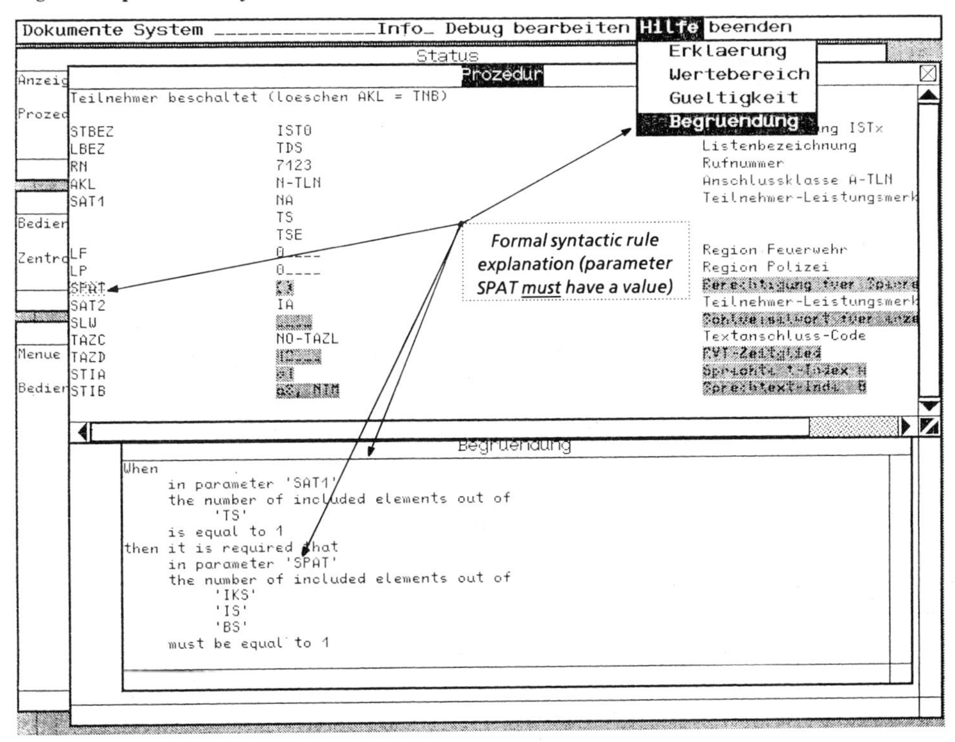

Fig. 5. Sample screen layout 2.

Finally, the executable code in the online expert operations system is produced using compiling techniques, converting formal logic rules into the conventional programming language.

Siemens-Albis has implemented the above functions (a), and (b) successfully in its advanced workstation PC-X20 running the UNIX operating system. Currently, first systems are being deployed in various PTT locations in Switzerland helping to maintain in the high grade of service within EWSD networks. Figures 4 and 5 give state-of-the-art examples for the operator's screen layout based on window system and *comprehensive guidance information* directly derived from the knowledge representation database. Conventional error-prone interpretation of the "paper" manuals is no longer necessary.

REFERENCES

CCITT, 1988. CCITT Rec. Z.301–Z.341. Man–Machine Language MML, Geneva.

Ribbeck, G. and Skaperda, N., 1984. EWSD as a basis for ISDN. In: Proc. ISS'84, Florence, May.